A DoD CCRP/NDU Collaboration

This collaborative effort is a continuation of the series of publications produced by the Center for Advanced Concepts and Technology (ACT), which was created as a "skunk works" with funding provided by the Assistant Secretary of Defense (C3I). The early success of ACT led to the creation of ACTIS when the president of the National Defense University (NDU) merged the experimental School of Information Warfare and Strategy with ACT and ASD (C3I) made the Director of ACTIS the executive agent for the DoD Command and Control Research Program (CCRP). ACTIS has demonstrated the importance of having a research program focused on the national security implications of the Information Age and in providing the theoretical foundations for providing DoD with information superiority, as well as the importance of an educational program designed to acquaint senior military personnel and civilians with these emerging issues. As a result, ACTIS's educational programs are being merged with the Colleges of NDU and ACTIS's research programs are being transitioned to OSD under the direction of ASD (C3I).

CCRP Publication Series

DoD C4ISR Cooperative Research Program

Library of Congress Cataloging-in-Publication Data

Krygiel, Annette J.
Behind the Wizard's curtain : an integration environment for a system of systems / Annette J. Krygiel.
p. cm. -- (CCRP publication series)
Includes bibliographical references.
ISBN 1-57906-018-8
1. Military telecommunication--United States. 2. United States--Armed Forces--Communication systems. 3. Internetworking (Telecommunication) 4. Expert systems (Computer science)
I. Title.
UG593.K79 1999
355.6'884--dc21 98-32110
July 1999 CIP

The National Defense University

The Institute for National Strategic Studies (INSS) is a major component of the National Defense University (NDU) that operates under the supervision of the President of NDU. It conducts strategic studies for the Secretary of Defense, Chairman of the Joint Chiefs of Staff, and unified commanders in chief; supports national strategic components of NDU academic programs; and provides outreach to other governmental agencies and the broader national security community.

Behind the Wizard's Curtain:

An Integration Environment for a System of Systems

Annette J. Krygiel

A Collaborative Effort Between the National Defense University and the DoD CCRP

CCRP Publication Series

Table of Contents

List of Figures

Acknowledgments

The Honorable Emmett Paige, Jr., the former Assistant Secretary of Defense for Command, Control, Communications, and Intelligence, has had a powerful interest in systems integration. When I first broached the prospect of my leaving a senior managerial role to investigate large-scale systems integration, he provided strong encouragement and offered many suggestions, among them an assignment to the National Defense University. What was then germinating in my mind as a seed quickly took stronger root. When I sought the advice of the Honorable John Deutch, for whom I had worked since he was Deputy Secretary of Defense, he endorsed the National Defense University as an institution of excellence. He made my conversion to an academic happen when Lt.Gen. Ervin J. Rokke (USAF), then President of the University, extended his hospitality. To these three I owe the opportunity that resulted in this book.

My home at the National Defense University began and ended in the Institute for National Strategic Studies, directed by Dr. Hans Binnendjik. It was he who piqued my interest in the TF XXI experiment when he suggested that I accompany him on a trip to view the U.S. Army's endeavors at Fort Hood, Texas. Dr. David S. Alberts has given unstinting and outstanding support to this book. He not only provided the necessary sponsorship in his role as manager of the C4ISR

Cooperative Research Program, but was available when needed, reviewed many drafts tirelessly, and provided positive suggestions constantly.

My investigation began by reviewing how the private sector was approaching integration on a large scale. A number of companies active in the defense sector offered both senior managers and engineers as participants in informal discussions with me in a (somewhat) organized agenda. Topics included methodology, best practices, trends in architecture and technology, resource implications, partnerships, customer roles, and acquisition practices. I am indebted to the following companies for the generosity of their time and their insights: Booz-Allen & Hamilton Inc., Computer Sciences Corporation, Raytheon Systems Company (IGS), Harris Corporation, Hughes Aircraft Information Systems, IBM, Intergraph, Lockheed Martin Corporation, The Mitre Corporation, MRJ, Rational Software Corporation, Sun Microsystems Federal, Inc., and TASC.

I am grateful also to many participants of the Digital Production System program and the TF XXI program who engaged in many dialogues with me, provided information, contributed documentation, answered many questions, and freely offered their own perspectives of the programs. Many of these individuals were asked on repeated occasions to clarify conflicting data, amplify existing material with details, and provide information when none was available from existing sources.

My reviewers deserve a special acknowledgement. As busy people, they often accomplished the reading of my many versions on their own time. Using their expertise and many suggestions, I was able to make this a better and more readable book.

Special recognition is given to Lois Burke of the NDU Graphics Department and to John McDaniel of Evidence Based Research, Inc., for their graphics support. Karina Braszo of the NDU Graphics Department was the creator of the magical cover. Thanks go to Dr. Richard E. Hayes and his staff at Evidence Based Research, Inc., for their instrumental support in transforming the book into its final edited form.

The last and ultimate words of acknowledgment are for the National Imagery and Mapping Agency (NIMA). Although established as recently as October 1996, it has been my parent organization for most of my career. Today it retains the vestiges of the Air Force's Aeronautical Chart and Information Center, and those of the Defense Mapping Agency and the Central Imagery Office, both former Department of Defense agencies. Collectively these organizations comprised 37 years of my professional life. Through NIMA's support of my assignment to the National Defense University, I was able to write this book. Today and in the future, NIMA faces the challenge of integrating the United States Imagery and Geospatial Information System. It is my hope that this work provides practical assistance for its team **behind the Wizard's curtain**.

While many sources were used to construct an understanding of the events that occurred during the two program ventures that comprise the case studies, the conclusions and recommendations are mine alone. I am responsible for errors of fact or interpretation, and the opinions expressed in this book are solely my own.

Preface

Twenty-first century concepts of operations that promise to transform us into an information-age military are beginning to emerge from each of the services and the joint military community. These concepts share a common vision, articulated in Joint Vision 2010, of being able to generate greatly increased combat power by creating and leveraging information superiority to achieve shared awareness, increased speed of command, a higher tempo of operations, greater lethality, increased survivability, and improved synchronization. The concepts also share a common assumption: the infostructure necessary to support these operational concepts will be there when needed.

To transform these emerging concepts into real operational capabilities will require much work to migrate the current collection of legacy systems infused and augmented by new and emerging capabilities into a coherent infostructure. Lack of interoperability and security are arguably the two most critical areas of shortfall. They are among our most vexing and persistent problems, and constitute the two largest obstacles we need to overcome. In some cases we will be able to exert the kind of leadership and impose the kind of standardization required to forge a system of systems and achieve the level of holistic behavior that is implicit in this term. In other cases we will find that we need to take a federated approach to developing the necessary degree of coherence. The extent to which a system of systems approach will be practical and the degree of success we will achieve with a federated approach remain to be seen. Our success will depend in large part on the degree of delivered interoperability and security.

This will not be an easy journey; it will be easier if we, as a community, develop a better understanding of the challenges involved, the problems encountered previously, and the actions needed for success. This is not just a technical problem that systems developers can solve. The problem encompasses moving beyond a collection of largely independent and stove-piped systems to a coherent infostructure capable of supporting emerging operational concepts. The task begins with how we think about enterprise processes and the information necessary to support them, and not just with individual systems. It includes how we develop future architectures and design and build participating systems to generate and use information in support of various missions and tasks. It extends to how we make infostructure investment decisions and how we think about achieving desired behaviors.

As Dr. Krygiel points out, there will be many mission-oriented systems of systems or federations of systems that share common information, computing, and communications resources. For each one of these it will be the end-to-end flow of information from various sources to its many uses that will need to drive the train rather than just the component processes along the way. The distinctions between the business side of the

Department of Defense and its warfighting side already are beginning to disappear as more information is made available to warfighters in a timely manner. This is truly a fundamental shift in the way we think about and do things today and it will require cultural change. Such significant changes will occur only if preceded by widespread understanding and a vigorous program of experimentation. Here, too, is the rub. It is hard to do the kind of discovery experiments necessary to really explore the opportunities to do business differently unless you can imagine interactions that cannot take place today. It is hard to imagine what is possible without some hands-on experience, but to get this necessary experience requires at least a critical mass of the systems of systems be available. We must ensure that our efforts at experimentation are well supported by an experimental infostructure.

Dr. Krygiel has performed a valuable service to the community by analyzing two very significant programs, each of which involved a system of systems. Her focus is on their integration efforts to extract the nuggets we need to develop a better understanding of the tasks ahead. We need to work together to ensure that these lessons recorded become lessons learned. As will be clear to her readers, there is still much to do and more to learn and understand about developing and fielding an effective and durable infostructure as a foundation for the twenty-first century. Without successfully fielding systems of systems, we will not be able to implement emerging concepts in adaptive and agile command and control, nor will we reap the potential benefits of network centric warfare. The C4ISR Cooperative Research Program (CCRP) will continue to pursue this line of research and to seek out collaborators to improve our understanding of this important area.

David S. Alberts
Director, Research OASD (C3I)

Introduction

Our U.S. defense strategy seeks new levels of effectiveness by harnessing the power of advanced technologies, particularly those of information technologies. A central premise to future military strategy is the formation of a system of systems (SOS) to attain dominant battlespace knowledge. By coalescing data from collection and processing systems, the resulting information can be integrated with systems of weaponry and warriors for a seamless sensor-to-shooter flow. Linking these with the capabilities of maneuver, strike, logistics, and protection will allow decision makers at every level to respond significantly faster than any adversary and in any operational situation.

Because a SOS is necessary to realize such powerful effects, I believe the integration environment to form such an entity should merit considerable focus. Having been a program manager for the integration of a relatively small number of individual systems, I find the achievement of this much larger venture a daunting challenge, and not one to which I would automatically ascribe success.

I had the opportunity to observe the appropriate level of attention on such processes as the U.S. Army labored to integrate hundreds of information systems to support a major warfighting experiment—Task

Force XXI (TF XXI). The Central Technical Support Facility at Fort Hood, where the integration was occurring, resembled a crisis operations center. The walls were papered with configuration item drawings, the boards covered with problem assessments. Engineers and developers rushed in a frenzy of activity to correct and change hardware and software. Meantime, hundreds of soldiers trained on the integrated product around the clock, occasionally expressing frustration with systems that did not work as intended or required. In short, it appeared so like the integration and training environment that had been used during my own program experience that I was immediately intrigued with the similarities.

Also analogous was the conviction demonstrated by the participants that the integration of multiple information systems is indeed challenging. Today in the era of the Internet and advertised interoperability, we are undermined by the expectation that properly engineered systems will plug-and-play and snap-in, snap-out. Instead integration is more typically a strenuous and complex undertaking to obtain the results intended. This is an era of chaos and non-linearity theory, and the relationships between networked information systems increasingly are described in terms of biological phenomena—unpredictable phenomena, at that.

This book then emerged from two case studies, the continuing analysis of the Army's TF XXI, and revisiting the Digital Production System (DPS) program with which I had been associated for many years while at the (then) Defense Mapping Agency.

This work is directed at recovering successful strategies and determining an environment that supports not only the integration of a SOS but its use for operational training. I characterize the activities that transpire as occurring **behind the Wizard's curtain**, a reference to L. Frank Baum's (1900, 1903) wonderful book, *The Wizard of Oz* . The reader may recall that the powerful wizard is finally revealed to the heroine Dorothy and her companions as a mere mortal, laboring behind the screen to produce the magic effects. This is an appropriate simile because it requires skilled people armed with modern arts to achieve the technological magic required to integrate and sustain a SOS. It is the people and the processes and infrastructure they use that this work examines.

Why my focus on the integration phase? The answer is many-faceted. It is true that the entire life cycle of a SOS merits attention, and this need is paramount to deliver the capabilities needed for the future defense strategy. More on this theme is covered in the section of this book on future work. The integration process and its associated environment do not provide an architecture or offer an alternative to a good design. So why emphasize integration? One reason is that it can be used effectively as an adjunct to the requirements and design processes if accomplished early enough in the life cycle. The U.S. Army demonstrated this with TF XXI.

I view the integration phase as the last certain opportunity to deliver an integrated product before its deployment for operations. Frequently it will be necessary to assemble such a capability despite inadequate previous processes—and using systems developed and operated for other and different purposes. Circumstances are not always optimum. Still, the integration process and environment, if sufficiently robust, can be used to overcome some disadvantages. They are certainly necessary to garner sufficient quality in the product no matter how robust the requirements, architecture, and design processes; otherwise, the operational community will suffer the impacts while attempting to accomplish its mission.

The Road Taken Through This Book

The road taken in exploring the topic of integration traverses many subjects: Joint Vision 2010, the defense enterprise, Greek philosophy, characteristics of a SOS, digital mapping, battlefield digitization, operational training, and recent experiences in Bosnia.

This book begins with a brief look at a SOS from the viewpoint of the U.S defense strategy and the framework of Joint Vision 2010. The intent is to provide an impression of SOS scope and the level of expectations for its operational support. There is not one SOS that is a one-time new start, as a single information system project. Rather a SOS more typically will be assembled from shared reusable components and with many existing systems independently developed for other and various missions.

Past problems with the interoperability of information systems have resulted in several strategies that are necessary and relevant to achieve the integration of a SOS. These are reviewed briefly, but the key question is whether they are sufficient of themselves. The answer is no. Current operations demonstrate this, as do the two case studies. The chapter also examines the nature of change and the consequences on a SOS. A principle is developed from the Greek philosopher Heraclitus: *"You can never experience the same SOS twice."* There is not one SOS but many SOSs. This arises from the need to use particular systems for different missions and the rate of change of circumstances and technology. As a result, the process of integrating a SOS occurs many times.

Chapter 2 poses the question: "What is a SOS?" Unfortunately there are no commonly accepted definitions. An answer is provided by building upon a few basic definitions. A SOS has constituent systems, which themselves are large-scale systems. There are different kinds of SOS, and the discussion of one type introduces the concept of a federation of systems (FOS). Coalition activities more characteristically require a FOS capability. The implications of a FOS in the Joint Vision 2010 era are not immediately assessed. Rather they are deferred until after the discussions of the SOS integrations accomplished during the two program ventures examined in this book.

The two case studies are the core of the book. Both programs, the DPS and TF XXI, produced an integrated product characterized as a SOS. Both were ventures that applied revolutionary changes in operational concepts through advanced information technology. Although different methodologies were used to develop their architectures, they followed similar strategies for integration. Their overviews are provided in chapter 3, and the integration experiences are described in detail in chapter 4. The two approaches are deliberately juxtaposed to illustrate the high degree of correlation, and there are more similarities than differences. Despite extensive preparation for integration, both the DMA and Army managers initially found themselves in real difficulties. Both recovered and moved on. Their experiences are worth telling and worth reading.

In chapter 5 the conclusions from this examination take the form of nine lessons learned about integration. These provide practical strategies for the team **behind the Wizard's curtain** to achieve a successful SOS integration. The lessons learned are fundamental and supplement good practices for program managers. Use of a structured integration environment in a single facility enables success—and with more effectiveness and efficiency. It also results in a more robust integrated product. Not the least of the lessons deals with the necessity for skilled people properly empowered, staffed, trained, and chartered within the acquisition process to accomplish the integration of a SOS.

Three additional lessons learned are described in chapter 6. These differ from those related to integration and apply to operational training on a SOS. Both DMA and the Army prepared their operators by training them for their missions on the SOS being integrated **behind the**

Wizard's curtain. What was demonstrated was the need for more and iterative training with a SOS. There is also a requirement to adapt training materials with considerably more exposition of the SOS capabilities, in addition to providing the operator with information about his or her individual system.

Considerable efforts were required to integrate the DPS and the TF XXI SOS, yet they are simple in comparison to the demands of the era of Joint Vision 2010. Because a FOS will be required for many missions, refinements or differences in the strategy for integration will be needed. The final chapter explores these topics and concludes that future experimentation should continue assessments of SOS and FOS activities **behind the Wizard's curtain**. Considering that future operations will be joint and coalition, integration will require processes characterized by increased collaboration. The required collection of systems will reflect a greater diversity. Processes will need to bridge many different cultures.

The lessons learned provide a foundation upon which to build. Several recommendations are provided in the final chapter, including the use of an integration environment. As defined, an integration environment describes who and what should appear **behind the Wizard's curtain**. Still, the more complex future demands continued investigation of other strategies and practices as well. Specific work is proposed including the analyses of more case studies. The extensive experimentation to implement Joint Vision 2010 planned for the decade ahead provides the opportunity for these additional assessments and investigations.

Chapter 1

The System of Systems

The U.S. Defense Strategy and the System of Systems

The National Security Strategy relies on the U.S. military to play an essential role in ways that protect and promote U.S. interests (The White House, 1998). Accordingly the future military strategy is proceeding in consonance with the concepts stated in *Joint Vision 2010* (*Joint Force Quarterly*, 1996; *Concept for Future Joint Operations, Expanding Joint Vision 2010,* 1997). Among the important precepts that comprise this powerful framework is that of information superiority:

> *—the capability to collect, process, and disseminate an uninterrupted flow of information while exploiting or denying an adversary's ability to do the same.*[1]

From a position of information dominance, the strategy seeks a powerful and seamless sensor-to-shooter flow. We will be in a vastly improved position to "see" our enemies, "decide" on a course of action, and

[1] As defined in *Joint Vision 2010.*

subsequently "destroy" or "influence," whichever is consistent with our national objectives. We will succeed by using information technology as a strong enabler for our decision-makers at every level and in every operational situation.

Central to this strategy is the formation of a "system of systems" (SOS) to achieve dominant battle space knowledge. This is a concept renewed and expanded by Admiral William Owens while serving as Vice Chairman of the Joint Chiefs of Staff (JCS) (Owens, 1996). He noted the superior technologies emerging in three areas:

- those of sensors for intelligence, surveillance, and reconnaissance (ISR)
- those computer processing capabilities supporting command, control, communications, and intelligence (C4I), and
- those surrounding and supporting precision weapons.

By coalescing the data from systems being developed in the collection and processing domains, we can realize a significant awareness and knowledge of the battlespace unable to be achieved previously. We can extend this strategy to integrate further with the systems of weaponry and warriors to achieve that seamless sensor-to-shooter flow so desired. And then, coupled with emerging technology, we can transform and link with the systems of maneuver, strike, logistics, and protection.

Among the essential components to implement this strategy is that of a capable network sufficient to support the aggregation of all these systems.[2] But ultimately, it is the *integration* of these many individual systems that will provide the means to collect, evaluate, and deliver the information needed to support the decision process and enable a significantly faster response.

Secretary of Defense William Cohen (1997), in building on the President's National Security Strategy, adopted the Joint Vision 2010 plan as the template for military operations of the future. Contained within Joint Vision 2010 are four operational concepts that, in the aggregate, are intended to provide the capability to dominate any adversary and control any situation. These are dominant maneuver, precision engagement, focused logistics, and full-dimensional protection. Certain key expectations about a SOS emerge from this framework and its powerful concepts, as follows:

- **Joint Vision 2010**—The future defense strategy requires that U.S. forces support a full spectrum of operations ranging from humanitarian assistance and peacekeeping to high intensity conflict. These vary in scale from small contingencies to major theater events of warfare. Coalition operations are essential to protect,

[2] A system is broadly characterized here as any sensor, platform, or weapon (in addition to the more traditional computer or network) through which bits flow and which can be connected to a network. As such, it can be embedded in a sensor and it can be strapped to a soldier. It can accommodate many purposes.

promote, and conduct our national interests internationally and to achieve global reach. Therefore collaborating with coalition allies—many of them in partnerships that are neither tested nor experienced before—becomes an important element of the strategy.

- **Dominant Maneuver**—Forces will be widely dispersed as joint air, land, sea, and space assets. There will be fewer overseas points from which to launch forces. Even today, forces have been migrating from a forward deployment toward one based in the continental United States. More extensive maneuverability will be required to accommodate force projections over strategic distances than ever before. Dispersion necessitates broader and more rapid collaborations from all assets and elements of the forces—and to mass greater effects. And rapid re-dispersion of forces is required to cover concurrent operational demands. This has implications on all levels, on tactical as well as on joint operations. Maneuverability also requires an evolution of organizations to become more agile and versatile.

- **Focused Logistics**—With the need for deploying and sustaining more expeditionary-like operations as well as to mass effects, the need to accurately locate, track, assess, and transport assets across geographic regions is important for agility. Information technology promises to achieve this without skewing the "tooth-to-tail" ratio unduly toward support. Tailoring logistics to the needs of a specific mission and positioning supplies as needed are both desirable objectives. Nevertheless, "just-in-time" provisioning must be balanced carefully against "just-in-case" provisioning, with a view to the consequences and risks.

- **Precision Engagement**—Precision engagement will be broadly applied. The future will provide more capable weaponry, real-time information of targets, situational awareness of friendly and unfriendly forces, more lethal munitions, and increasingly precise delivery against the objective. Future performance will be assessed by an external environment with expectations set at zero collateral damage and no fratricide. The capability to engage precisely and react accordingly will evolve to the level of an individual combatant; this requires a ubiquitous integration across all levels of an operation.

- **Full Dimensional Protection**—The future brings new threats, and therefore increased vulnerabilities. Consequently, there will be a need for increased protection. Collection assets of surveillance, reconnaissance, and intelligence will be used to detect and track attacks from conventional and unconventional means. New sensors for the characterization and detection of chemical and biological warfare agents are required to deal with these threats, increasingly used by rogue nations and terrorist groups. Protection of the forces is paramount. Protection of the information infrastructure,[3] upon which these future operational forces will rely so extensively, is also necessary.

Operational Expectations of a SOS

What do these strategies and concepts demand of a SOS? A few declarative statements synopsize the operational expectations:

- A SOS must support an operational tempo substantially increased from that of today.
- The individual systems of a SOS must be configured and linked to sustain both strategic and tactical level activities, accessed and applied by a theater force as well as a highly mobile expeditionary unit.
- A SOS must reach widely dispersed sites around the globe.
- A SOS must encompass the systems providing information superiority with those of command and control, with weaponry and engagement, and with those supporting maneuver, strike, protection, and logistics.
- A SOS must be appropriately secure and protected.

[3] The term "infrastructure" is broadly inclusive. It comprises the underlying computers, communications, organizational base, people, and processes to support the function specified by the context.

- A SOS must be *joint*, linking service and agency assets.
- A SOS must support coalition operations, connecting to the systems of international partners to the appropriate extent.

These expectations indicate a substantial scope for a SOS, and an integration of systems beyond anything previously attempted or achieved.

The Journey to the System of Systems

The SOS is not just a futuristic concept initiated by Joint Vision 2010 and the potential of information technology. In one sense, the Department of Defense (DoD) has been on a journey to evolve something like a SOS for some time, although not of such breadth, scale, coherence, or level of connectivity and interoperability.[4] It has long been recognized that exchanging information and sharing services between and among information systems is important. However, experiences during *Operations Desert Shield /Desert Storm* revealed serious deficiencies in the ability to do this. Shortfalls even resulted in the fratricide of friendly troops (Stanley, 1998).

In reaction to the severity of interoperability problems in the Desert War, a program, "C4I for the Warrior," was initiated by the JCS. This was structured into several

[4] Interoperability is the ability of two or more systems or components to exchange and use information (IEEE Standard 610.12, 1990).

phases and moved in the direction of applying a common set of standards to improve interoperability. The program also promoted a transition from military standards to commercial standards (Starr, 1996).

A defense enterprise architecture is being evolved. Standards and guidelines for the enterprise facilitate interoperability. An agreed-to subset of the standards comprises the current version of the Joint Technical Architecture (JTA)[5] (DoD Joint Technical Architecture, 1998). Compliance with the JTA is considered mandatory. With the multitude of systems being acquired, enhanced, and maintained, efforts have been increased to ensure compliance—including oversight of acquisitions, the establishment of a Chief Information Officer[6] in each defense organization, and the expansion of certification testing.

The growth of common services is also a fundamental strategy to foster interoperability within the defense enterprise. The Defense Information Infrastructure (DII) comprises networks, computers, software, data bases, applications, interfaces, and services, and it provides a common operating environment (COE). It incorporates key joint systems that provide significant functionality common for all users. Two prominent examples of these are the Global Command and Control System (GCCS) and the Defense Information System Network, both in

---

[5] The first version focused on C4I systems. The second version expanded into domains such as combat support, weapon systems, and modeling and simulation.

[6] This resulted from the Clinger–Cohen Act of 1996 (formerly the Information Technology Management Reform Act), PL 104–106.

operation. The GCCS is a C4I system that provides common services and a shared data environment, accessible by information systems that fit within the overall architecture (Butler, Diskin, Howes, and Jordan, 1996; DII Master Plan, 1998). When service-unique extensions of GCCS must be acquired, they comprise an adjunct to the joint, common capability. This contrasts with past practices of proceeding with separate and sometimes disparate acquisitions by the services and agencies.

The defense enterprise will evolve as its various components evolve. It provides an actual foundation for a future SOS. A SOS is rarely a one-time new start, but rather it will build on common capabilities such as the GCCS. A SOS will include legacy systems. It also will incorporate new capabilities, many specifically developed to implement the Joint Vision 2010.[7]

Legacy Systems and the SOS

A unique factor that compounds the challenge of assembling a SOS from existing systems is the sheer number of legacy systems in the DoD. These systems were developed without benefit of the enterprise architecture definition and before the JTA was defined.

[7] A task force identified at least 32 critical functional capabilities required for Joint Vision 2010 and conceived a future global SOS spanning all levels, strategic to tactical. The results were documented in the Advanced Battlespace Information System (ABIS) Task Force Report (1996). The roadmap for science and technology initially provided has since been refined to mesh the new with current capabilities to plan assimilation incrementally (Joint Warfighting Science and Technology Plan, 1998).

Many do not even use modern digital technology and were developed for standalone environments.

Using legacy systems to support missions will continue indefinitely. Defense downsizing, the enormous funding required for recapitalization, as well as the time required for migration all compound their duration. Some systems take as many as 10 to 15 years to acquire, and typically remain in the inventory between 20 to 40 years.[8] Legacy systems are challenging for the private sector as well, but the length of the acquisition cycle[9] and the number and age of systems in the inventory are probably unique to DoD.

With the rate of change in technology, today's developmental systems become tomorrow's legacy systems. They add difficulty to constructing a SOS because generally they are not compliant with the then-current version of the defense architecture. Interoperability with them is problematic. Differences must be reconciled with those of systems that fit within the overall enterprise framework. Legacy systems must be migrated to a state of compliance through enhancements.

[8] Former Secretary of Defense William Perry, at the 1997 Association for Computing Machinery (ACM) conference, "The Next 50 Years of Computing."

[9] DoD continues to streamline the acquisition process by moderating requirements for oversight, processes, and documentation based on the risk of a specific acquisition.

The Journey to a SOS—Emulating the Commercial Sector

The DoD strategy to achieve interoperability is adopting many processes and methods that have proven successful in the commercial information technology sector (Schaeffer, 1998). An architecture should be comprised of systems that are "open"[10] as well as modular, with various components that can be reused in future evolutions.

Emulation of successful business practices of the private sector has resulted in some streamlining of defense acquisition processes and wider application of commercial engineering, design, and development processes. It also has brought the greater use and leveraging of commercial technology and commercial services in lieu of specialized components and unique elements. While commercial products have all the merit of reduced development expenditures, these bring additional advantages for achieving openness, and, therefore, interoperability. There is evidence that the commercial sector of the information technology domain has been marching smartly toward open

---

[10] An open system is defined as *"a system that implements sufficient open specifications for interfaces, services, and supporting formats to enable properly engineered applications software: (1) to be ported with minimal changes across a wide range of systems, (2) to interoperate with other applications on local and remote systems, and (3) to interact with users in a style that facilitates user portability"* (DISA DII Master Plan, 1998).

architectures for some time[11] [12] because open systems provide economic benefits.

Consequently commercial products and services will constitute a larger percentage of the future defense architecture (and a SOS). Industry standards comprised more than 60 percent of those in the initial version of the JTA and nearly 70 percent of the second version. Subsequent versions increasingly will rely on commercial enterprises and international standards.

The Reality Check

With the current migration toward an enterprise, the emphasis on standardization, common components, and the expanded use of commercial products and services, the questions are: Are they sufficient to achieve a functioning SOS? And given the conceptual framework of Joint Vision 2010, will they remain sufficient?

What is learned from current operations such as in Bosnia and from joint and service exercises and demonstrations is that interoperability between systems and the integration of multiple information systems continues to be difficult. A good synopsis of problems

[11] In an interview with Kevin Kelly, John Hagel discussed his book *Net Gain* and noted that between 1985 and 1990 there was a huge redistribution of shareholder value between companies that were owners of proprietary architectures and those that were championing open architectures (Hagel, 1997).

[12] Kevin Kelly (1997) noted the phenomenon of open systems and common standards emerging to maximize the potential of network infrastructure as Rule 8 of the "New Rules of the New Economy."

and causes connected with the lack of interoperability is provided in the Commission on Physical Sciences, Mathematics, and Applications, "Realizing the Potential of C4I: Fundamental Challenges" (1999). The strategies followed, when viewed as a complete solution, are necessary but not sufficient. The case studies discussed within this work will illustrate this also.

Larry Wentz's summary of lessons derived from *Operation Joint Endeavor* in Bosnia provides insight into the challenges and difficulties of achieving a SOS (Wentz, 1998). To succeed with the integration of communications and information systems (CIS) to support the implementation force (IFOR), he recounts the following:

> *The challenge facing NATO and the nations was to build a long haul and regional CIS network out of a mixture of military and commercial equipment that would vary widely in age, standards, and technology and would be built very quickly once given the order to deploy. Putting the pieces of the puzzle together would most likely not result in a true 'system of systems' for IFOR. Furthermore, there would be a need to interface systems that had not been planned or designed for interfacing. The independent national systems would be tied together, not engineered as a single system. Given the uncertainty of the situation it would most likely be a case of integrating what you get, not necessarily what you need, and then making the best of it.* (Wentz, 280–282)

Other ventures outside of defense also indicate the challenge. Unlike Fermat's last theorem,[13] the proof of these difficulties is not left entirely as an exercise for the reader. There is a rich body of literature available on the challenge of succeeding (and failing) with complex information systems. The Standish Group (1995) undertook a survey called CHAOS on information technology projects in the commercial sector. Results were grim. Generally, more than 30 percent of projects were canceled before completion; more than 50 percent cost nearly twice their original estimates; and only 16 percent were completed on time and within budget.

Capers Jones has characterized the difficulties in succeeding with software intensive information projects as a function of their size, as characterized by function points (Jones 1996a, 1996b, 1996c). These provide units of measure for software size using logical functions as an alternative to lines of code.[14] Only 14 percent of projects that exceed 100,000 function points are delivered on time. Risk rises dramatically from the 10,000 to 100,000 function point range and higher.

Of the two case studies discussed in this book, one is equivalent to 100,000 function points, and the other to

[13] Pierre Fermat was a French mathematician who, in proposing a new assertion, said *"I have a truly marvelous demonstration of this proposition, which this margin is too narrow to contain."* Three and one half centuries lapsed before anyone could discover the demonstration.

[14] To do rigorous function point analysis of a software application requires inputs, outputs, inquiries, logical files, and interfaces.

10 times that.[15] The two programs are more complex than the software project of a single system, but this simple comparison provides a lower bound on difficulty. This theme is developed later in this work. And while these endeavors are challenging, even more comprehensive capabilities will be required to support the Joint Vision 2010 era.

Learning about the System of Systems from Heraclitus

The Greek philosopher Heraclitus observed in the fifth century B.C. that change was continuous (de Laguna, 1921). His philosophy is epitomized by the line:

You can never step into the same river twice.

The Joint Vision 2010 era must deal with continuous change. Many capabilities in a SOS are founded on information technology, which itself is an accelerating phenomenon.[16]

Adapting to the fast pace of technological change has significant implications. Accommodating change itself becomes a primary requirement for a SOS.

Technological innovation is not limited to friendly forces, but rather it is proliferated globally. High-tech

[15] Based on lines of code estimates of approximately 10,000,000 for DPS and 30,000,000–40,000,000 for TF XXI (and using 100 lines of code per function point).

[16] For example, W. Brian Arthur compared the rate of technology evolution to biology evolution. He concluded that: *"...technology is evolving at roughly 10 million times the speed of natural evolution"*(Arthur, 1997).

weaponry is available in the marketplaces of the world, as are commercial communications, navigation, and transportation assets. The net result is a dynamic situation with respect to the capabilities of adversaries and, therefore, the overall threat.

Asymmetric threats to the United States can arise through the manipulation of information technology by unscrupulous nations or groups with otherwise inferior assets. While technological superiority is one cornerstone of the national military strategy, faster technology insertion is needed for the United States to retain its competitive edge. Accommodating rapid technological turnover has become a strategic necessity.

The pace of innovation also will require faster adaptation of the enterprise framework and an increased pace for migration of legacy systems. Strategies to deal with rapid change such as the reliance on commercial technology are anticipated to ease the accommodation. However, the revolution in the information technology teaches that increased capabilities lead to increased requirements and increased complexity. The Chief Technology Officer of Microsoft observed that software is limited only by human expectations:

> *The software crisis is perpetual because the benefits of panacea solutions are absorbed by rising expectations. The real driver is expectations.*[17]

[17] "The Next 50 Years of Software," by Dr. Nathan Myrhvold, 1997 Association for Computing Machinary conference, "The Next 50 Years of Computing."

The global environments also are changing rapidly. With an increasingly wider set of players on the world's stage, more diverse geopolitical environments are anticipated. Alterations in economic, technical, societal, religious, cultural, and physical conditions will occur. The U.S. defense strategy is based on the reality of living in a dangerous, uncertain, and unpredictable world where rogue nations can use asymmetrical strategies such as terrorism and weapons of mass destruction with devastating consequences. The net result of these circumstances is flux, a spiral of altering operational concepts, revised strategies and doctrine, and organizational adaptations, all continuously accommodating diverse and exceptional circumstances in a changing world. These changes, in turn coupled with technological innovation, drive the need for more adaptation of capabilities. These circumstances are interrelated, linked, and coevolved. The net result is a state characterized by continuous change.

Considering the circumstances surrounding the SOS, a "Heraclitan principle" is derived by paraphrasing:

You can never experience the same SOS twice.

Different Systems of Systems for Different Missions

This Heraclitan principle is confirmed when considering that, as in the past, the future will require that U.S. forces engage in distinct operations with different combinations of systems. These differences arise based on the specific type of mission, the nature of the operational environment, the duration of the mission, as well as many other factors. Many components do remain the same, but others vary, some unique for the geopolitical environment. Humanitarian assistance in Somalia requires certain capabilities that are different than those for a desert war in Iraq.

Peacekeeping is a mission different than that of conflict and involves a significantly different approach to command and control and decision making. *Operation Joint Endeavor* is characterized as an operation other than war (OOTW). Mission requirements for such operations, when contrasted with those of major conflicts like *Operations Desert Shield/ Desert Storm*, demand different combinations of systems.

The accounts of peacekeeping in Bosnia are illustrative. For *Operation Joint Endeavor*, a unique situation arose because NATO was out of its normal area, and because a large number of countries that had never worked together collaborated. These circumstances lead to an extraordinary command and control structure, and consequently to extraordinary combinations of command and control systems (Layton, 1998).

Weather and time of day are also among factors that contribute to differences in capabilities for different missions. They establish conditions whereby one reconnaissance information system is more advantageous than another. For example, electro-optical imaging sensors cannot penetrate through clouds or the night so that other types of sensors are used. Terrain is a factor that distinguishes the selection of surveillance capabilities. In Bosnia, extensive terrain masking and inadequate resolution initially precluded extensive use of the Joint Surveillance Target Attack Radar System (JSTARS), which was better adapted to detecting opposing wartime force movements rather than distinguishing intertwined friendly and unfriendly forces. Later when there was increased freedom of movement, it was used for tracking military vehicles in conjunction with other assets (Wentz, p.102).

Participants in actual operations noted these variations in suitability for OOTW missions:

> *Systems that work in deserts may be useless in jungles, forests, or urban centers. Tools that are safe in open areas may have unacceptable consequences in crowded areas. Where the immediate threat is low, technologies that work slowly or require detailed preparation are useful, but they cannot help in urgent situations.*
>
> *Complicating the process further is the fact that technical requirements vary with the location, type of operation, and the time available for application.... In* Desert Storm, *for example, soldiers reported that they could spot buried mines using night vision devices. While this worked in that desert environment, it does not work in forest or jungle areas. Likewise, technologies that work in fields may not work in hills and probably won't work in urban environments.* (Alberts, 1995)

"Stepping into a SOS" could be an experience of very short duration. A particular SOS may be configured and used for a period of days or weeks to support a mission-transient operation. Other combinations of systems may be integrated and sustained for longer periods of time. During that time individual components or systems may undergo adaptation, proceeding through many versions. For prolonged operations, such as in Bosnia, even the common framework capabilities may evolve. Also, new advanced capabilities may be introduced at any time to provide an operational advantage.

The U.S. forces are expected to conduct multiple operations concurrently. This implies sustaining various combinations of systems concurrently.

Behind the Wizard's Curtain

Considerable attention will remain focused on the implementation of Joint Vision 2010 in the decade ahead. Changes to personnel, organizations, and doctrine will be evaluated and addressed. Advances in technology will be used to achieve the capabilities required. The SOS so central to the defense strategy must support the operational expectations. However, the SOS is an *integrated* product for operational missions. An integration process will be needed, not just one time and not just for a single product. There will be many integrations, and many combinations of systems integrated, and a continuum of integrations, depending on the different objectives intended and the duration of missions.

To support each operational activity that must play on center stage, another production also must play. That production is the orchestration of people, processes, and the information systems **behind the Wizard's curtain** to provide the technological magic so necessary for the mission's success. While these activities are not at the center and front of the stage, they are essential.

It can be argued that with sufficient prescience, anticipation, warning, and resources, the Wizard's team will deliver the required SOS. A modicum of integration of systems was achieved for *Operations Desert Shield/Desert Storm*, although many participants would consider this a stretch. In Bosnia, with sufficient lead-time, many systems were integrated, including many with advanced technologies. On the other hand, by imposing conventional limits on resources and time, or without sufficient warning, the rapid integration of large numbers of systems is well beyond our current abilities to achieve today. Compounding the difficulty is the fact that the nature of today's threats results in responses that often combine forces and systems in unanticipated ways. Among the most formidable challenges to realizing the promise inherent in the future defense strategy is the integration of information systems, a process based on interoperability. It merits considerable attention.

Chapter 2

Systems of Systems and Federations of Systems

What is a system of systems (SOS)? This short chapter develops an answer by building upon a few basic definitions. There are distinctions between a single system and a SOS. A common understanding is necessary to appreciate the descriptions and analyses of the two case studies provided in subsequent chapters.

The concept of a federation of systems also is discussed. It is representative of the capabilities required for the Joint Vision 2010 era and its discussion provides a context for integration methods needed for the future.

The terms and definitions are listed in Appendix D, Glossary of Terms.

Systems and Systems of Systems

The concept of a "system of systems" has been used in various sciences for many years. A 1964 paper on New York City refers to *"cities as systems within systems of cities"* (Berry, 1964). The social, biological, as well as the physical sciences make use of the concept. However, as applied to information systems, there is no widely accepted definition or agreement on how a SOS differs from more conventional systems.

The term "system" has more universal acceptance. Eberhardt Rechtin and Mark Maier (1997), who have developed the art of architecting systems, have provided a definition which would be recognizable in many sciences:

> *A system is defined as a set of different elements so connected or related as to perform a unique function not performable by the elements alone.*

This description conveys that a system has some essential ingredients—components, and relationships, and implicitly a boundary, that separates it from the rest of the environment. For an information system, the elements and components refer to hardware, software, and even people. The relationships are its interfaces, interrelationships between and among software components and hardware components, or between a user and any or all of these.[1] There can be interfaces to the external environment as well.

A system produces results unachievable by the components alone. Rechtin offers a general example to illustrate this aspect of a system, worth repeating here:

> *...imagine that your automobile was completely disassembled and laid out on your driveway. All*

[1] A more formal definition of interface is "shared boundaries across which information is passed" (IEEE Standard 610.12, 1990).

> *the elements individually would be just as before, all in working order. But you would have no transportation. Transportation, the unique system function, only exists when all the elements are connected together and function as a whole.* (Rechtin, 1991)

But what is a SOS? There is no commonly accepted definition, and there are differing classifications of large complex systems as SOSs (Shenhar, 1994; Eisner, 1993; Maier, 1996, 1998). To illustrate the inconsistency, at least one large defense system has been both characterized and disavowed as a SOS.[2] Both case studies analyzed in this book—the Defense Mapping Agency's[3] Digital Production System and the U.S. Army's TF XXI—are characterized here as a SOS.

In this book, the more conventional definition of a single system is expanded to that for a SOS:

> *A system of systems is a set of different systems so connected or related as to produce results unachievable by the individual systems alone.*

Characteristics of a System of Systems

Maier (1996, 1998) has provided characteristics to distinguish a SOS from more conventional systems. Certain of his descriptions are useful in the context of the two case studies and for considering the needs of future operations in the Joint Vision 2010 era—many systems managed by many organizations integrated to

[2] Global Command and Control System

[3] Now the National Imagery and Mapping Agency

achieve results to meet the objectives of a specific mission. He notes that in a SOS the various components are large-scale systems in their own right.

- Each is capable of independent action and fulfills a purpose of its own.
- The individual systems of the set are managed independently—to fulfill their stated purposes.

In contrast to Rechtin's example of an automobile given earlier, in a SOS the individual entities would not remain laid out on the driveway until assembled. They are capable of independent action. These constituents fulfill purposes of their own and can operate when disassembled from the whole. They are managed for their own purposes.

As for any system, there are links in a SOS as well. These are between the individual systems of the set and are the interface relationships necessary to accomplish those objectives not able to be obtained by the individual constituents alone.

Figure 2-1 illustrates the use of these terms for a single system and a SOS.

In addition to the two characteristics discussed, Maier (1996) provides three others that are useful to distinguish a SOS from the more conventional system:

- A SOS manifests emergent behavior because it achieves purposes not resident in the individual constituents.

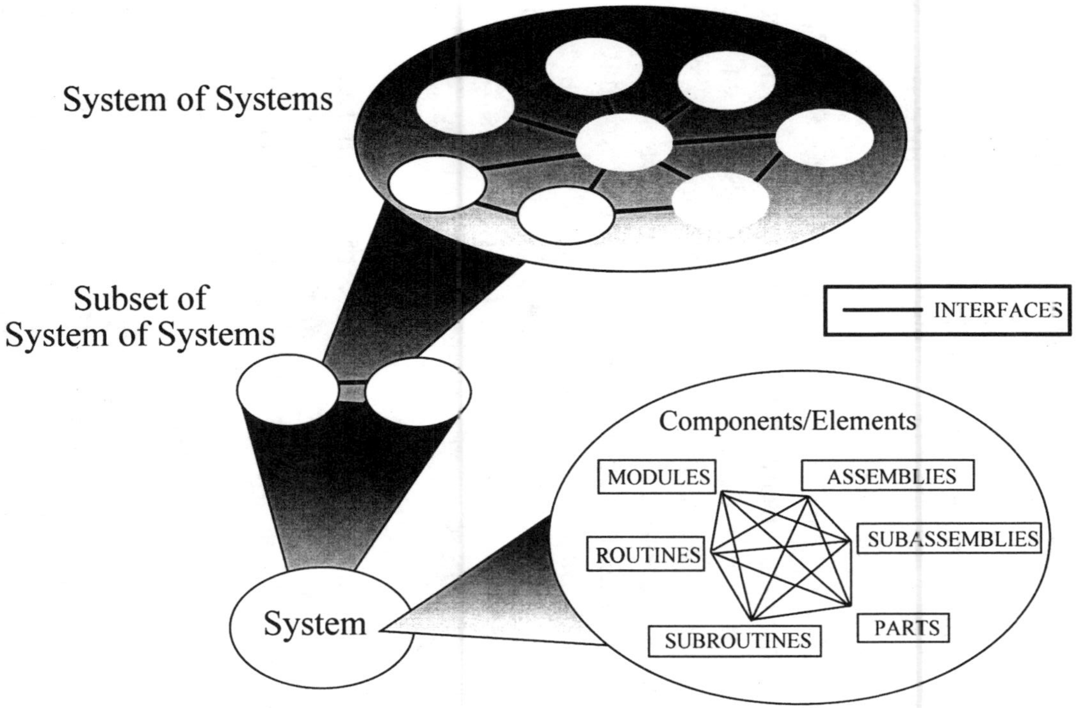

Figure 2-1. Hierarchy of a System of Systems

- It is evolutionary—functions are added, removed, and modified over time.
- It usually is distributed on a large geographic scale.

Prominent examples of a SOS cited[4] are the Internet, integrated air defense networks, and intelligent transportation systems.

The Internet provides a simple example of a SOS. Its constituents are many computer networks and major computer nodes distributed globally. Some of the individual networks may be further decomposed, such as into other subnetworks and computer systems. Internet nodes collaboratively exchange information using protocols. The Internet evolves with phenomenal growth. It also exhibits emergent behavior, typified by the complex distributed applications on its communications layer, including that of the World Wide Web.

Maier's examples also illustrate the point that the constituent systems of a SOS can themselves be a SOS, such as the World Wide Web, or a corporate enterprise.

Interoperability and Integration

A SOS carries out a purpose separate from those of its individual systems—through the relationships of its constituent systems. The interface relationships between and among the systems of the SOS achieve

[4] Maier (1996, 1998) provides discussions of these examples in his papers.

synergy—through the passage of information. Therefore, interoperability is an essential requirement for a SOS:

> *Interoperability connotes the ability of two or more systems or components to exchange and use information.* (IEEE Standard 610.12, 1990)

Interoperability enables the relationships, and integration ensures that the synergy of the individual systems realizes the purpose of a SOS. The integration event unifies the individual systems of a SOS to achieve the desired holistic behavior—to deliver the required results. Integration is as essential as interoperability for a SOS.

Usually the process of integration is both an incremental and a cyclic one—not just one of plug-and-play, although having it be so would be the ultimate objective. It is one of build and test, iteratively refining a system's components and interfaces until the required purpose is produced. For a SOS, it is building and testing the individual systems and their interfaces to determine if the result, the integrated product, meets the operational requirements. As shall be seen from the narration of the integration events of each case study, a great deal of effort was necessary to ensure each SOS delivered the results intended.

The Architecture of the SOS

The integration event of a SOS could be the milestone event comically characterized by "a miracle occurs here"—if it occurs without the proper system architecting and engineering preceding it. While integration is the focus of this book, the architectural framework and engineering efforts are relevant. In the overview of the two case studies provided in the next chapter, background on their architectures is provided as pertinent information.

The architecture of a system is its organizational structure.[5] It can be described using multiple and various perspectives, called viewpoints, each of which provides different information. The descriptions of the two case studies apply the same three[6] viewpoints typically used in describing the defense enterprise—operational, technical, and system architectures. As a general characterization, the systems architecture is developed using the standards described in the technical architecture to meet the requirements of the operational architecture.

Operational Architecture

Generally the operational architecture captures what the user expects to do and what information will be needed and exchanged by the organizational units. For the DMA case, the user was the Agency's workforce, primarily cartographers. For the Army's TF XXI program, it was a brigade of soldiers equipped with new digital capabilities. Descriptions of the operational architecture communicate the user functions, the

[5] As defined in IEEE Standard 610.12, 1990.

[6] Rechtin and Maier (pp. 120–122) present six viewpoints in their book, including a separate information (data) viewpoint. In contrast, the information model is embedded within the three viewpoints used in this book.

information required, operational relationships, and, if known, performance bounds:

> *A description (often graphical) of the operational elements, assigned tasks, and information flows required to support the warfighter. It defines the type of information, the frequency of exchange, and what tasks are supported by these information exchanges.* (DISA DII Master Plan, 1998)

Technical Architecture

The technical architecture describes the standards and guidelines with which the system must comply, such as information, processing, and transport protocols:

> *...the services, interfaces, standards, and their relationships. It provides the technical guidelines for implementation of systems upon which engineering specifications are based, common building blocks are built, and product lines are developed.* (DISA DII Master Plan 1998)

As later described in the two case studies in this book, DMA used very broad guidelines for its architecture, while the Army developed a detailed technical architecture.

Systems Architecture

The systems architecture provides a physical view and describes the real system components, as built and implemented, i.e., a description of:

> *...the physical connection, location, and identification of key nodes, circuits, networks, warfighting platforms, etc. and specifies system and component performance parameters.... The systems architecture shows how multiple systems within a subject area link and interoperate, and may describe the internal constructions or operations of particular systems within the architecture.* (DISA DII Master Plan 1998)

Most of the constituent systems of DMA's SOS were new developments initiated at the same point in time. In contrast, the Army's SOS incorporated many existing systems in addition to new initiatives. The systems descriptions of these legacies were an important starting point for developing the architecture for the TF XXI.

Federations of Systems

Each of the two case studies produced a SOS. As will be apparent in subsequent chapters, these were strenuous undertakings. Yet each was simple in comparison to integrating a SOS that matches the operational expectations for the Joint Vision 2010 era. One reason is that both programs had centralized management and authority that controlled the development and subsequent operations. Generally, there was control exercised over the three architectural viewpoints—operational, technical, and systems.

The Joint Vision 2010 era includes missions that will require a SOS when there is not the same centralized control and authority. Coalition operations will involve dispersion of power and authority and introduce many viewpoints from coalition partners. At the same time partnerships will bring greater diversity of assets than was experienced in either case study. This type of SOS is distinguished here by the term, "federation of systems."

A federation of systems (FOS) is a SOS—but one managed without central authority and direction. The constituent systems of a FOS are independently managed, and have a purpose of their own. But the degree of management independence is much greater. Power and authority are decentralized in management, development, and operations. Because there is no central power or authority for direction, the participation of the constituents occurs through collaboration and cooperation to meet the objectives of the federation. Consequently a FOS is generally characterized by a greater degree of *autonomy, heterogeneity,* and *distribution.*

This concept of a FOS borrows from other work,[7] such as the taxonomy introduced for federated databases by Amit Sheth and James Larson (1990) and from collaborative structures by Maier (1998). It also dovetails with the principles of federations provided by Charles Handy (1992), albeit these were developed in the context of organizations. Handy defined five

[7] The intent is not to present a complete taxonomy, but to discuss characteristics of federations to develop an understanding of the complexity of integration events in the future. The interested reader should consult the references.

principles[8] of a federation, the most important of which is subsidiarity, the assignment of power at the lowest possible point in the organization. In such a structure, each element has great autonomy in participating to meet the overall objectives of the federation.

This relationship of a SOS to a FOS is captured in figure 2-2. This figure is *notional* and indicates the *relative* positions of a SOS from a FOS. There is no algorithm provided to define precisely the demarcation. Nor are there values on the axes to delineate any boundaries.

The region labeled SOS is characterized by (more) centralized management and control over development and operations, and therefore less autonomy. While the systems of the SOS are managed independently and have a purpose of their own, as a constituent in the SOS they are subject to some form of direction. The result is (more) uniformity in architectural framework, guidelines, standards, and development principles, and a (more) uniform operational view than would be the case for a FOS.

A SOS for a coalition operation, such as in Bosnia, can be viewed as positioned farther along the autonomy axis into the FOS region. A SOS supporting service or agency missions is closer to its origin. A SOS for a

[8] The other four principles are: interdependence to spread power and avoid risk; a uniform way of doing business; separation of powers to keep management, monitoring, and governance in segregated units; and twin citizenship to ensure a federal presence in an independent region (Handy, 1992).

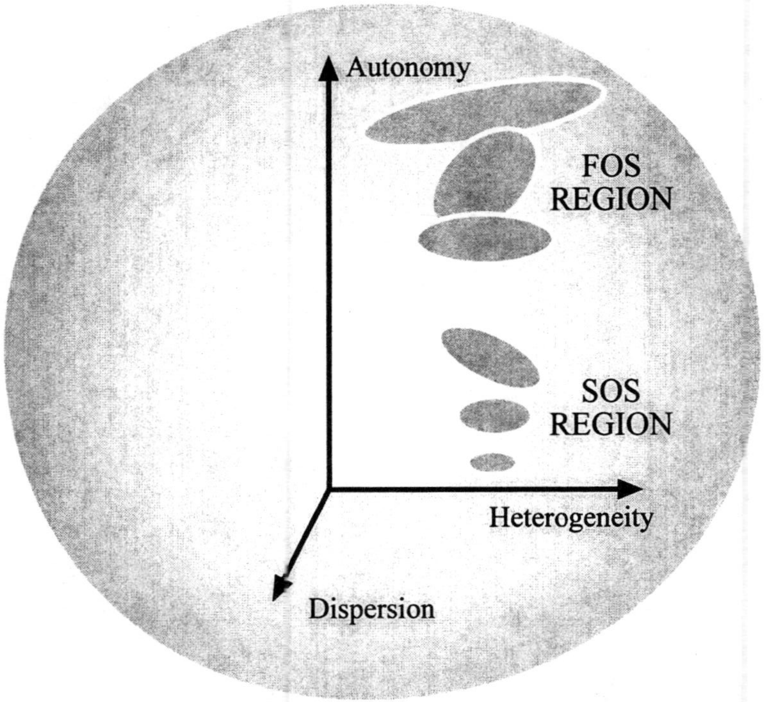

Figure 2-2. System of Systems (SOS) and Federation of Systems (FOS)

joint mission (not shown) would be depicted between the two regions.

Figure 2-2 also depicts different shapes for each SOS and FOS to convey different degrees of heterogeneity and dispersion. Each could vary by degree depending on the nature and complement of the systems supporting the needs of a particular mission. The degree (and shape) could change with evolution of the SOS or FOS—as systems are added or altered. Also, a specific FOS could be less dispersed geographically than a specific SOS.

The attributes of autonomy, heterogeneity, and distribution are interrelated. As power shifts from central direction to more collaborative management of development and operations, the characteristics can change by degree. As control is decentralized and localized, then local requirements and local interpretations can become more diversified. Local power with autonomy contributes to variations in requirements and therefore system capabilities supporting these. Similarly, greater geographic dispersion can contribute to increased autonomy, which results in more localization of processes, concepts, and even cultures. These, in turn, can lead to singular requirements, which in turn leads to more heterogeneity.

To return to the example given earlier, the Internet began as a SOS under the management and direction of the Department of Defense. Today it is a FOS. Little central management or power of enforcement exists. Collaboration by participants is voluntary. Today a footprint of its constituent systems and characteristics would be very different than one from its early years.

The Implications of a Federation of Systems

Generally a FOS is more heterogeneous than a SOS. Any SOS will exhibit *some* diversity, even one adapted to a single architectural framework. The individual systems are managed and operated independently by various organizations (e.g., services and defense agencies). When multiple communities of humans manage, engineer, and operate, there will be different interpretations of requirements, standards, and priorities. The legacy systems in the defense enterprise already have been flagged as amplifying heterogeneity.

The technical architecture of a FOS could comprise widely varying standards and guidelines—the equivalent of multiple technical architectures used by its constituents. Accordingly, the individual systems as implemented would be (more) heterogeneous than a SOS in hardware components, platforms, operating systems, programming languages, software applications, and data structures. There would be more issues of interoperability.

The coalition operational environments will bring multiple cultures and multiple languages, as well as different rules, guidelines, processes, values, and constraints. The rich mix of international players will introduce interoperability issues of semantics—disparate meanings, and different interpretations, in addition to different languages. There will be differing management, development, and operational environments. A FOS will be comprised of diverse individual systems managed by collaborating organizations but developed and operated using their own methods, processes, and technologies.

The operational viewpoints, when as diverse as in coalition operations, result in individual systems in a FOS that mirror the varieties of operational approaches and relationships in their individual architectures. Decentralized control implies variations in how operational communities perform their missions—and at what level. This affects how their information systems are developed and operated. For example, command and control might be exercised differently by various organizations in a coalition operation. Such differences would be reflected in the individual systems of the FOS that are used by the various partners.

From every architectural perspective—operational, system, and technical—for a FOS the interoperability issues will be greater and the integration more challenging than that of a SOS. Nonetheless, when it is possible to simplify architecture and relationships among the constituents, such as in the case of the Internet, phenomenal synergy is possible.

Hard, Harder, and Hardest

To an extent, the term "system of systems" is an unsatisfactory one. It masks the complexity and difficulty of the integration challenge. The unitary nature of the term screens the multiplicity, diversity, and autonomy of the individual systems and their management communities. However, it is powerful in conveying the synergy of integration.

While integrating a single large scale system is hard, the integration of a set of systems independently developed and operated for distinct missions is harder, and a FOS even more so. As such, these differences have resource, schedule, performance as well as management implications.

The approaches for integrating a SOS will be examined through the experiences of the two programs that comprise the case studies. Each is characterized as a SOS, but one (the TF XXI case) is positioned relatively closer to the FOS domain than the other. They provide a good starting point for determining how to succeed with a SOS integration.

It may be inferred that the approach and procedures used to integrate the components of a single system are sufficient for success. They provide a good basis, but in dealing with a SOS, refinements to methods and processes are required for activities **behind the Wizard's curtain**.

It can be anticipated that a FOS integration requires yet still another evolution of strategy from that of a SOS because a FOS is developed with diverse architectures and managed through collaboration rather than direction. The intent is to examine the implications of the FOS in the final chapter of this work by building upon the lessons learned for SOS integration developed in earlier chapters.

Chapter 3

The Two Case Studies

This chapter provides an overview of the two programs with which this book is concerned—the Defense Mapping Agency's[1] Digital Production System and the U.S. Army's Task Force XXI. They share many characteristics in common, not the least of which was their goal to deliver a SOS with revolutionary capabilities.

Because the emphasis of this book is on the integration environment, the architecture, engineering, and development aspects of the two ventures are discussed only to the degree necessary. But these *are* relevant. Many bibliographic references also are provided which can be used by the interested reader for more information.

Management structures for the programs are discussed because they are particularly germane to the integration.

[1] Now the National Imagery and Mapping Agency.

Defense Mapping Agency's Digital Production System

The Digital Production System (DPS) was a 10-year development by the Defense Mapping Agency (DMA) to deliver an end-to-end digital processing pipeline for production of mapping, charting, and geodesy (MC&G)[2] products. It was conceived with a sense of urgency in the early 1980s and resulted in one of the then-largest development programs[3] undertaken in the Department of Defense (DoD).

Its genesis was a series of studies, many congressionally sponsored, to look at collection platforms for the 1990s in the context of emerging requirements, particularly for weapons systems. Its birth was the direct result of one of those studies—the Hermann Panel Report.[4] This report recommended that DMA expedite development of a modernized production line that accommodated digital softcopy source materials and used computer-assisted techniques. Dr. Richard DeLauer, then Under Secretary of Defense for Research and Engineering (USDR&E), directed DMA in February 1982 simply

---

[2] MC&G comprises *"the collection, transformation, generation, dissemination, and storing of geodetic, geomagnetic, gravimetric, aeronautical, topographic, hydrographic, cultural, and toponymic data"* (USIGS Glossary, 1998).

[3] An advisory board made this program assessment just before the DPS critical design review. Board members observed that while the scope of the software and complexity of the integration effort made it comparable to major special programs within the Department of Defense, there was no comparable production program.

[4] The panel was chaired by Dr. Robert Hermann.

to implement the Hermann report. DMA delivered the full operational capability (FOC) in November 1992.

The program was considered critical to the success of the defense mission. At the time there were significant backlogs of requirements for MC&G products while there were growing dependencies on them, particularly in weapons systems. In addition there were requirements for products providing greater fidelity of the earth's terrain and culture to drive simulators to support rehearsals for military missions. While the panel professed skepticism about the specificity of particular requirements, it concluded that the requirements were paramount and growing.

Weapon systems were increasingly dependent on descriptions of terrain and cultural features to aid their navigational capabilities. The digital data products prepared by DMA were used for smart weapons, as well as for the air-based, sea-based, and land-based missiles. As an added impetus, there was also great reliance on DMA processes to provide targeting information with increasing precision.

The Hermann panel anticipated future scenarios that were even more time-sensitive than then-current demands, requiring a crisis-like turnaround response from DMA. The members saw the steady trend of increasing precision with respect to the relative and absolute accuracy of the positions of points and features on the earth, although this would vary depending on the nature of the product. With some prescience (this was 7 years before the end of the Cold War), the panel anticipated the growing unpredictability of the locations of crisis events. This situation, when viewed in the aggregate with the increasing reliability on MC&G data, gave urgency to the need to produce products more accurately and in a more timely manner. Extraordinary efforts were expended by DMA to meet crisis demands,[5] which at the time, depending on the product need, could require many months of activity.

By the early 1980s DMA long had been regarded as the world's premier map-making organization. While the conversion to digital products was incorporated into the Agency's longer term strategy, the production plants with the installed and aging technology base relied largely on film-based source materials and a hybrid of analog and digital processes. At the time, DMA's ability to enhance the speed of its production processes and the accuracy of its products relied on a modestly endowed research and development program. Most of the production systems within the organization were stratified, fragmented by processes tied directly to the nature and formats of the source materials used, as well as by the type of individual products required. Changes in the formats of source materials or in customers' product requirements inevitably resulted in adaptations that introduced further delays in meeting requirements.

[5] Examples were DMA's support to the hostages' rescue in Iran and operations in Africa and the Middle East.

Need for Revolutionary Technology and Revolutionary Concepts

In assessing the long-range requirements in concert with emerging collection and processing technologies, the Hermann panel perceived a chance to achieve a significant breakthrough. The opportunity was marked to position for increased precision, adapt for crisis scenarios, and also accommodate the growing need for MC&G data. But the modernization of DMA into digital processes needed significant investment, a faster-paced research and development program, and a large acquisition venture. The panel anticipated significant advances from automation to substantially reduce the timelines, particularly those involving the correlation of stereo imagery, critical to the derivation of particular DMA products.

Exciting possibilities from technology were envisioned. At the time, some of DMA's most labor-intensive (and therefore time-intensive) processes included using stereo imagery to extract elevation and feature data. Pipeline times of 2 to 3 years were not unusual to produce new products from source materials. It was believed that delivery times could be substantially reduced with the introduction of more automated processing techniques.

The Hermann panel was not concerned with impacts on production resources as a result of such significant change, nor even expected that reduced resources for production processes would result. In fact, the panel did not consider this aspect to any great extent. It noted that the impact of accommodating new materials was probably underestimated, and probably not understood very well. But as the modernization program proceeded, DMA put tremendous effort into reducing production inefficiencies. The design approach adopted to do this had a significant impact on the overall success of the program.

The Hermann panel concluded its report in 1982 with recommendations to USDR&E to proceed with substantial investments in creating a new production capability at DMA using digital, time-responsive processing and based on digital source materials. Anticipating the extensive scope of the venture, the panel further recommended the establishment of a separate organization responsible for the modernization and acquisition of substantial equipment and systems, and that DMA develop a plan to accomplish this.

Responsible Organization Established

DMA responded to USDR&E by establishing a Special Program Office for Exploitation Modernization (SPOEM) in February 1982. Despite no previous experience with a development of this magnitude, the Agency proceeded aggressively with the program called the Digital Production System (DPS) under the strong leadership of a program director specifically appointed to the task. By design, a single organization and a single senior manager were designated as accountable for the acquisition and development.

The Defense Mapping Agency Program—in Two Phases

The program was partitioned into two phases—an interim phase called Mark 85 and the final phase called Mark 90. Mark 85 was so-named because the deliveries were to begin arriving in 3 years—in 1985. Analogously, Mark 90 derived is name because the initial operating capability (IOC) was to occur in 1990.[6]

From today's perspective, a 10-year development might be judged as non-time critical. However, there was a great sense of urgency to get on with the conversion to digital materials and develop a new digital production infrastructure. The venture was assessed as an enormous undertaking with a great deal of technical risk and uncertainty. At one time DMA had conceived that such a capability would require at least 15 years to achieve. But DMA, in reaction to the DeLauer letter, determined that the early 1990s provided a reasonable delivery schedule. DMA moderated expectations by early introduction of the understanding that only several years after the program's final delivery of hardware and software would a full production capacity be achieved.[7]

---

[6] The IOC schedule was slipped a year to 1991 because of consequences of the Balanced Budget and Emergency Deficit Control Act of 1985, 2 USC 901 (Gramm–Rudman–Hollings Law).

[7] The design relied on population of a data base of MC&G data, which when reaching sufficient detail over geographic regions, could be used to generate or revise products more efficiently. Because this initial compilation of data required several years, full production throughput was also delayed.

Mark 85, an Incremental Step

As initially conceived, Mark 85 had as its objectives a delivered capability to ingest and exploit new source materials using *film*-based processes. The strategy was fixed on retaining many of the then-existing production systems retrofitted with new software and hardware. Mark 85 capitalized on some work that had occurred in previous research and development efforts by the Agency and incorporated several softcopy techniques and digital components that were enhancements in capability. Some of these provided important insights for Mark 90 developers.

Primarily the approach emphasized enhancements of then-current processes. It was an evolutionary step—not a revolutionary one. As a result, the operational concept retained some of the disadvantages of the fragmented, product-specific processes and standalone computer processing systems, all of which were connected primarily through manual processes. One new addition was that of an automated production management capability, which became heritage to the Mark 90.

Importantly, Mark 85 was also a risk-mitigation implementation. By delivering this interim capability, DMA ensured that there would be a production capacity in place long before the more ambitious second phase was ready. This interim approach turned out to be critical and enormously successful, especially when unanticipated operational requirements exploded in

Operations Desert Shield/Desert Storm before the delivery of the Mark 90.

Mark 90, a Revolutionary Step

This book focuses on the Mark 90 phase of the DPS; it was truly revolutionary in its conception. It used hitherto untried digital source materials. It went beyond the state of the art in extraction of geospatial[8] data using entirely softcopy techniques. It developed digital automated processes for cartography akin to those of image processing. At its inception, it required advances in data management to support volumes of imagery and geospatial data and enormous bandwidth in data communications to move that information locally and to geographically dispersed sites.

In the early 1980s, the notion of storing MC&G data in a digital database and using that as a base from which to generate a variety of cartographic products was revolutionary. As an overall strategy, it promised great flexibility for adaptability and tailoring of product information to accommodate the plethora of emerging weapons systems, as well as increased currency of information through more rapid revision. Such capability had never before been implemented and only conceptually articulated.

Production processes would access imagery that was 100 to 1,000 times greater than the average size of a

---

[8] This includes *"information that identifies the geographic location and characteristics of natural or constructed features and borders of the earth"* (USIGS Glossary, 1998).

digital data set in use at the time, and produce a unit of MC&G data that was about 10 times greater. As a result, the concept of accessing, exploiting, and disseminating multiple images and extracting and storing MC&G data went well beyond the available technology and methodology of the time. At the outset[9] of the program in 1982, only 10 percent of the technology required for the DPS was commercially available.

In addition to its innovations in use of digital imagery and digital processes, the DPS was based on the strategy of multiproduct operations. The cartographer extracted information for many products at the workstation in one job assignment, rather than extracting information for one product over multiple job assignments at different points in time. Optimization of this time and labor intensive process promised a breakthrough in the overall annual production rate of Agency products.

The Demise of a Prototype

At one point in the program history there was a Mark 87, but it had an early demise.[10] Its elimination was barely noticed at the time, but it later resulted in serious complications for Mark 90, evident in the integration phase with which this book is concerned. Its purpose was the delivery of a fully integrated digital prototype—an engineering model—less ambitious in performance

[9] By IOC in 1991, approximately 90 percent of the technology was commercially available. Some became available from vendors who developed commercial versions of the Mark 90 technology.

[10] Its termination as a development was announced in September 1983 at a concept design review status briefing.

than the total Mark 90, but offering full functionality and including all individual systems.

Because the prototype could not be completed with sufficient lead-time to influence the manufacturing, its benefits were considered marginal. At the time of its termination, it was believed that other means could be used to verify the DPS engineering.

The Digital Production System Architecture

As the DPS undertaking was evolved, it was partitioned into aggregates of functionality termed "segments" in the program terminology, but here called "systems" consistent with the definitions provided earlier. The Mark 85 consisted of six systems that were principally hardware and software enhancements augmenting then-existing systems or ongoing development initiatives. Later two of these, the Hardcopy Extraction System and the Source Acquisition System, were further modified and became legacy systems included in the set of systems of the Mark 90. A substantial heritage of software modules and some hardware components from the Data Integration System of Mark 85 were subsumed into the Mark 90 system.

Mark 90 was the DPS system of systems (SOS). General overarching requirements for functionality, performance,[11] and specific MC&G products were partitioned and aggregated into separate parts. These,

[11] The DMA established objectives of 75 percent reduction in production time and 50 percent reduction in production costs for products generated in the DPS.

in turn, were used for the acquisitions of five individual systems and the three heritage Mark 85 systems already mentioned. When integrated and delivered, the Mark 90 SOS had seven individual systems interfaced through complex relationships between and among the individual systems.

The DPS was designed to produce 24[12] MC&G products, primarily from digital imagery but also using varieties of other information including foreign-produced maps and charts, film, and text. While the 24 products constituted a small number of the hundreds of products produced for customers, at the time they collectively required a majority of the Agency's resources to produce.

Figure 3-1 illustrates the DPS architecture of seven constituent systems as a pipeline. Based on requirements for MC&G products from users, a production program was developed. Production managers then used many factors to determine the subsequent flow of work and information. These included the availability of imagery (if unavailable, it was acquired), the preparation and georeferencing of source materials for the assignment, subsequent extraction by cartographers of terrain and feature information, and eventually generation of graphic products, both hardcopy and digital. Source imagery and source materials, the extracted MC&G data, and varieties of intermediate data were stored, accessed,

[12] The number of products varied during the program as a result of evolving requirements, priorities, and the efficiencies of the digital processes. At FOC there were 24 products.

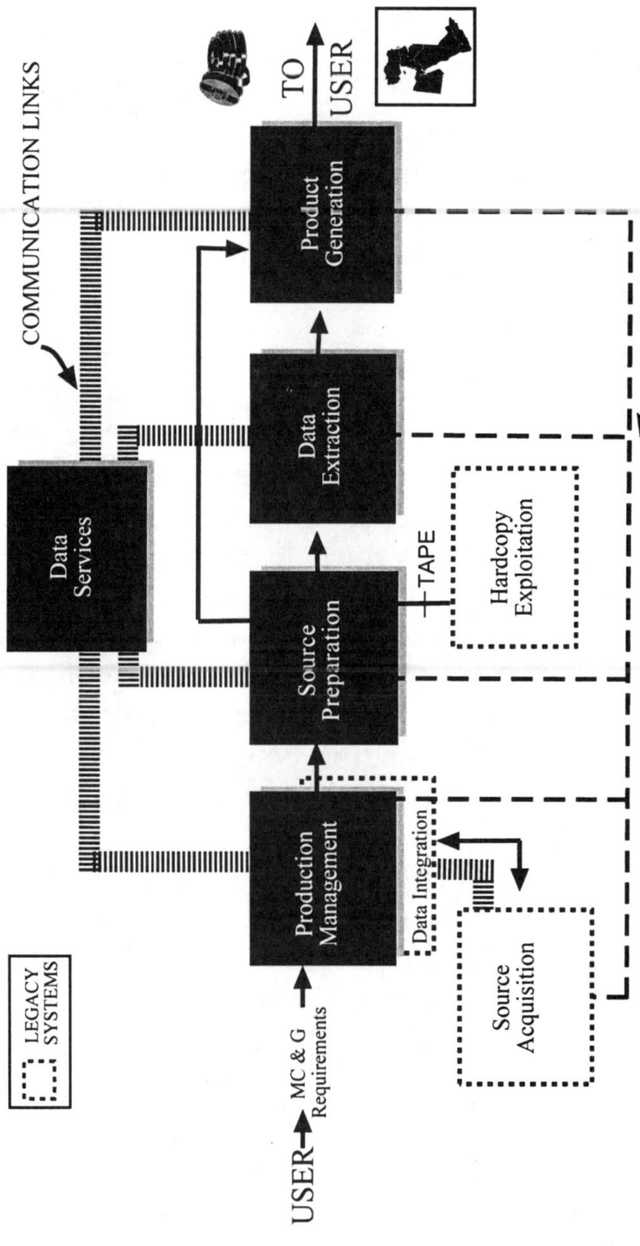

Figure 3-1. Digital Production System

and disseminated at many points through the pipeline by using a set of data and communications services.

The DPS was installed at three geographically dispersed Production Centers. At the time of delivery, the user community of DPS was the major subset of the Agency's production workforce and numbered about 3,000 professionals. They were in organizations which were structured along the lines of the pipelined production process.

The operational concept for the DPS gradually evolved from an early one of relatively independent actions by the various user organizations to one that relied on a great deal of coordination and communication to deliver the intended results—a set of mapping products derived from imagery, from other sources of information, and (ultimately) from a data base of existing geospatial information. This became true for users and organizations within a Production Center, for those coordinating assignments and information between Centers, as well as for those in the Headquarters. The principal factor contributing to this change was the centralization of authority in production management and its increasing role in scheduling and allocation of jobs, people, and equipment. Over time this resulted in a DPS structure characterized as tightly coupled,[13] with interdependencies among the constituent systems that complicated the integration challenge and ultimately adaptability. At the time it was believed that this would lead to significant improvements in production throughput and reductions in the numbers of people required.

The systems architecture "as built" for FOC was extensive, with large numbers of hardware and software components developed and delivered for the DPS. More than 3,000 pieces of equipment were acquired, including 2,000 workstations, about a thousand of which were based on developments specialized to the mapping processes. About 10 million lines of code were generated, integrated, and tested, including nearly 2 million knowledge-based rules.[14] As installed, the DPS required more than 380,000 square feet of facility space at the Agency's three Production Centers.

Program Methodology and Schedule

The DPS was developed using a classic waterfall methodology (Winston Royce, 1970). From general requirements for the SOS, aggregates of functionality were partitioned into individual acquisitions. These resulting systems of the DPS became aggressive developments by different companies begun at about

[13] Two systems are coupled if they are interdependent (i.e., if at least one system requires information from the other, or requires components, services, or people). Tighter coupling indicates greater (i.e., multiple) interdependencies between systems than does loose coupling.

[14] Heuristics implemented in software to increase the degree of automated assistance to the cartographer.

the same time (with the exception of the Mark 85 legacy systems). After the allocations of functionality were initially made for the individual systems, there were few changes in the segmentation.

The DPS architecture evolved principally bottom-up, through the increasingly detailed functionality of the individual systems and the growing specificity of the relationships[15] between them. There was no equivalent of the defense enterprise common operating environment or the joint technical architecture available at the time. There were a few key standards that were applicable, particularly those relevant to customers' product requirements for MC&G data.

As implemented, the various systems of the DPS were heterogeneous, the diversity among them amplified by the nature of the special developments in each of them. The unique hardware and software components ranged in technology from image processing to cartographic generalization software to network switches. The very nature of the graphic MC&G products required developments specialized for the mapping applications, such as printers, scanners, and plotters with large size formats. Differences in the missions of the individual Production Centers required variations on components, equipment, and system configurations.

Designs were constrained by a general framework of programming practices and conventions and the

[15] As an example of complexity, one interface document required approximately 10,000 pages to express the details of the information exchanges that one system had with the others.

standards applicable at the time. By the time of FOC, 15 different programming languages had been used. Standardization was realized more within the components of an individual system than across the SOS. However, there were partial successes in limiting the numbers of platforms and operating systems through common mainframes and minicomputer environments. Multiple commercial products were incorporated, primarily for the information processing and data base environments, and some for communications.

Figure 3-2 provides a simplified DPS program schedule—requirements and design reviews, integration events, IOC, and FOC. At all the major DPS milestones there were assessments to examine the SOS architecture while considering the state of the individual systems and the interfaces between and among them. Modeling, analyses, and simulations were used to examine the viability of achieving functionality and performance, and issues were identified and addressed.

The integration phase occurred after completion of the development and testing of the individual systems and their interfaces. The SOS integration approach used a series of formal demonstrations to verify the correctness of the interfaces between systems of the set. These events included production-like jobs to generate MC&G products. Informal integration activities began even earlier. Discussions on the details of the integration events will be provided in the next chapter.

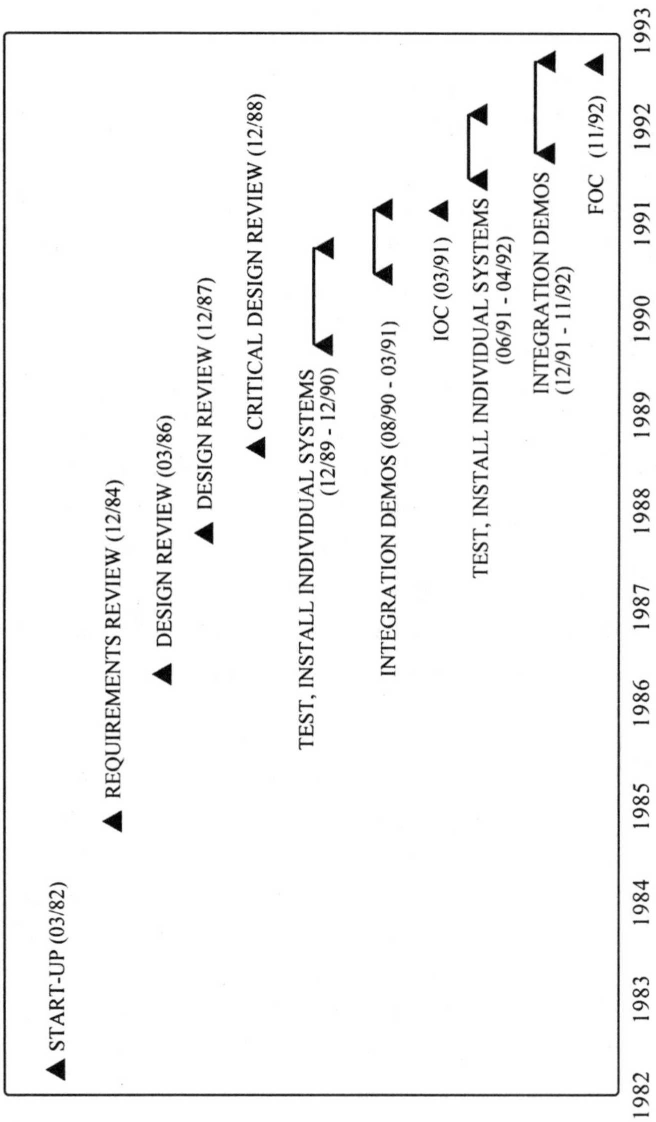

Figure 3-2. DPS Schedule

The DPS was delivered to a single site for the IOC. This delivery had reduced functionality, a decision forced by late closure of all the interface relationships, primarily those for production management. A similar orchestration of activities occurred before the FOC delivery, which included full functionality of the SOS. The individual systems and interfaces were completed and tested, then reintegrated as a SOS using another and different series of demonstrations at a single site. Then the integrated product was incrementally delivered to all three Centers. After achievement of FOC, the business of using the DPS for production and population of the MC&G data base began.[16] As DMA had planned, residual discrepancies were addressed while production usage gradually increased.

After Full Operational Capability

The FOC occurred in November 1992, 10 years after the initiation of the program. By the time of delivery, the customers' needs had dramatically altered and continued to do so over subsequent years, driven primarily by changes in the geopolitical environment. In response, the Agency migrated from a posture of producing MC&G products toward one of providing geospatial information and services. While the DPS is

[16] A caveat was that the actual use of DPS for production began *before* FOC. The production processes using DPS were serialized so that after intermediate stages of the pipeline were judged ready for production, the turnover to production occurred incrementally.

used for production today, it represents only a subset of the Agency's current systems architecture.

The DPS was not used as conceived because the needs and production scenarios had changed by the time of its delivery. Its design precluded easy and rapid adaptation. Alternative production capabilities were spun from certain DPS technologies, some in response to *Operations Desert Shield/Desert Storm*, which occurred before FOC; others in response to different customer requirements. A good general reference describing the DPS and its subsequent evolution is found in (Littlefield 1995). This paper also provides insight as to the effect of the DPS development on the commercial availability of systems and workstations—for cartographic processes and feature extraction. The state-of-the-art of commercial cartographic technology advanced by building upon the DPS development investments.

The overall DPS performance, which required a stable production program, a populated data base, and a strategy of multiproduct extraction (conditions never fully realized), was never achieved. Yet even today, for certain production processes, it surpasses even current commercially available capabilities.

With the DPS the DMA achieved the key objective it was given: to establish digital softcopy processes to produce MC&G products from digital source materials.

The Management Structure

The DPS was a development managed by a single Agency and intended for its own internal operational use. Figure 3-3 depicts the organizational structure used to manage the development and its transition to production. The responsibility for the acquisition and development of the DPS resided with its Special Program Office for Exploitation Modernization (SPOEM), later re-organized as the DMA Systems Center. One Agency Program Executive Officer (PEO) was accountable and reported directly to the Director, DMA. This individual was in a position to resolve any programmatic, engineering, and funding issues attendant to the acquisitions, including any connected with the integration.

At peak activity, as many as 400 DMA people worked on aspects of the integration. These included the SPOEM's program managers with their development teams, and the operational community leading the transition processes and participating in the acquisitions. One organizational element, called the "SOS cadre" in this book, was responsible for the DPS SOS. Among their responsibilities was leading the integration event with system integration and system engineering contractor resources supporting.

DMA's operational community participated throughout the program. To prepare for the acceptance of the DPS and its turnover to production, an Activation Control Team (ACT) was established in June 1989. The members played a significant role during the integration.

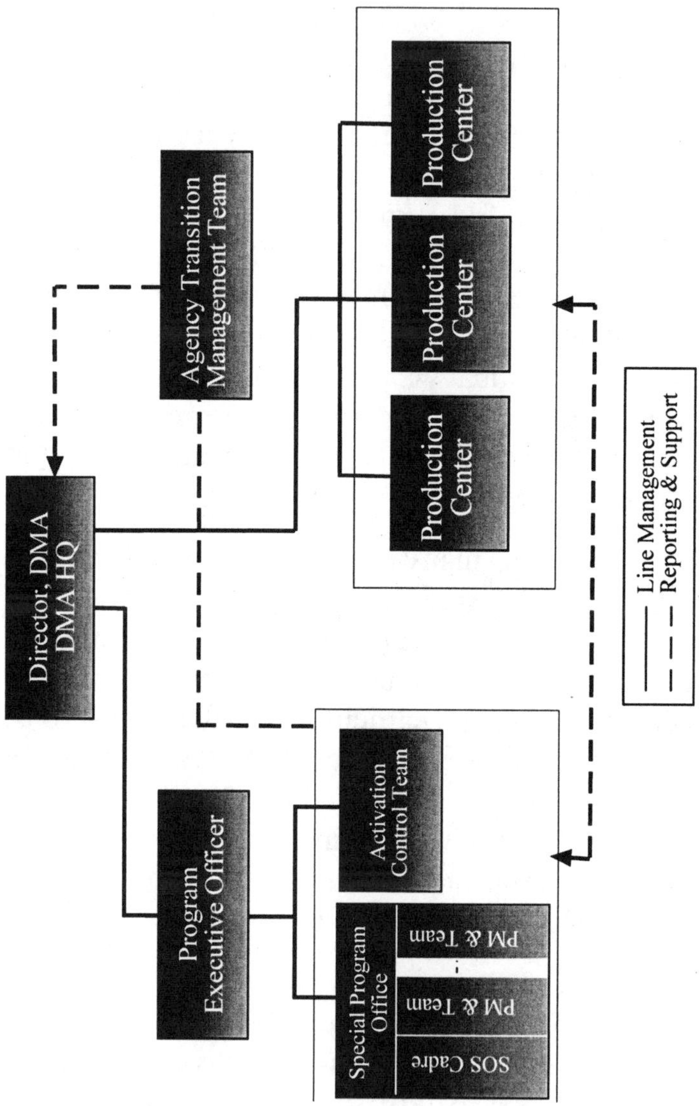

Figure 3-3. DPS Management Structure

Their leader reported to the PEO but was dual-hatted as the Director of the Production Center where the SOS integration occurred. He provided a focused voice for the operational community. The Director of DMA also convened an advisory team of senior leaders from the Agency's organizations, both production and acquisition, named the Agency Transition Management Team (ATMT). This group was chartered to ensure that an integrated planning and implementation approach was taken for integration, verification, and production ramp-up activities.

The U.S. Army's Task Force XXI

The Task Force XXI (TF XXI) resulted from a significant movement toward change in the U.S. Army. In 1992 the (then) Army Chief of Staff General Gordon Sullivan recognized that the convergence of several factors called for a significantly altered strategy for the Army. The geopolitical environment had changed as a result of the end of the Cold War. The U.S. Army was being downsized and based primarily in the United States. But the threats were becoming unpredictable, not only in their nature, but also in their location. Also information technology was becoming increasingly available.

General Sullivan concluded that the Army needed to shift toward a strategy of force projection with a demonstrated ability to rapidly alert, deploy, and conduct operations anywhere in the world (Sullivan & Dubik, 1994; Sullivan & Coroalles,1995). He recognized that information technology could provide a key enabler. He expected fundamental changes in every aspect of the Army, including structure, doctrine, capabilities, training, and tactics. However, the nature of these would require time and effort to understand.

A sequence of exercises and experiments called the Louisiana Maneuvers was conceived as a kind of laboratory to learn about the Army of the 21st century (Sullivan, 1992). Over the next 2 years much internal examination occurred. Many activities such as major exercises and simulations were used to assess the Army's ability to meet different force projection scenarios. What emerged among the conclusions was the importance of information technology and the opportunity to organize around information to mass effects.

The results of the Louisiana Maneuvers were far-reaching. Three complementary processes were initiated—the re-design of the operational Army, the re-design of the institutional Army, and the promotion of information age technology. This third component the Army called "digitization," and it was to be facilitated by a newly created Army Digitization Office (ADO). The TF XXI program and processes then emerged with the objective of transforming the current Army to one organized around information and information technologies for the 21st century.

The Task Force XXI Plan and Beyond

The Army subsequently defined a comprehensive program for battlefield digitization. The goal was:

> *to apply information technologies to acquire, exchange, and employ timely digital information tailored to the needs of each decider, shooter, and supporter* (*Providing the Means,* 1994)

By providing the communications and processing capability to influence speed, space, and time within the battlespace, two key advantages could be gained: shared situational awareness[17] and enhanced battle command. Implicitly such a strategy relies on the ability to accomplish integration between information systems; consequently, the ADO's mission statement included the coordination of integration.

The execution of the TF XXI plan was to provide the understanding of how to evolve. The Army needed decisions on doctrine, structure, and capabilities by the year 2000 to field the "Army XXI"[18] after that date. The TF XXI events would be used to evaluate needed changes. Subsequent evolution would result in "the Army after Next," a longer term objective, anticipating

---

[17] Defined as: *"the ability of a unit to know where its friends are located, where the enemy is, and to share that information with other friends, both horizontally and vertically, in near real-time"* (*Providing the Means,* 1994).

[18] The Army XXI program includes fielding a digitized division by 2000 and a digitized corps by 2004 (Reimer, 1998).

a significantly different force with greater strategic and operational mobility (Reimer, 1996, 1997).

A key element of the TF XXI plan was to use the strategy of an Advanced Warfighting Experiment (AWE). The magnitude of changes for the Army XXI necessitated an entire series of experiments, each successively building upon lessons learned from the previous. The AWEs were unique in providing the operational conditions to focus the technological developments, innovative operational concepts, and new force structure with experimental doctrine for evaluation. They, in turn, also built upon previous technology demonstrations.

The Army embarked on battlefield digitization using a bottom-up approach for experimentation, echelon by echelon, involving multiple systems and digitization initiatives. Early efforts began at the company level. In April 1994 the Army conducted the first of a series of large-scale exercises applying digital information technologies—one at the battalion level.

The Desert Hammer AWE took place at the National Training Center (NTC) at Fort Irwin, CA. Some key lessons derived from the experiment affected preparations for the TF XXI AWE, specifically the need to prepare extensively for fielding of and training with experimental systems. Over the next 3 years, exercises and AWEs were used to evolve concepts, doctrine, and capabilities. These included Prairie Warrior/Mobile Strike Force, Roving Sands Theater Missile Defense, Focused Dispatch, and Warrior Focus.

Task Force XXI Advanced Warfighting Experiment

The TF XXI AWE consisted of a series of live and constructive simulations that began in March 1996. It culminated at the NTC during 2 weeks of March 1997 with a major force-on-force encounter between the opposing force (OPFOR) and an experimental brigade trained and fully equipped with all the capabilities the Army's digitization program could provide at the time. This event required a SOS, the second case study of this work.

The essence of the AWE was to examine the effects of digitization on lethality, survivability, sustainability, and operational tempo. New doctrine was developed for the experiment and organizational changes were made in order to examine the effects of these changes when the experiment was conducted.

The brigade selected as the experimental force (EXFOR) was the 1st Brigade from the Fourth Infantry Division (4ID) based in Fort Hood, TX. The EXFOR was a brigade task force of 5,000 soldiers. It was comprised of three battalions of mechanized infantry, light infantry, and armor, with supporting field artillery, aviation, and engineering elements, and a reconnaissance troop (Hanna, 1997). The preparation of the EXFOR began in January 1996 and continued 24 hours a day until equipment deployment to the NTC in December 1996. Planning the experiment and defining and engineering its architecture began much earlier, at least as early as January 1995.

A simplified schedule for the TF XXI, highlighting the evolution of the SOS architecture, is shown in figure 3-4. The Army identified particular key core digitization capabilities needed for the field experiment, but was also willing to consider additional initiatives and prototypes that might provide a significant advantage on the battlefield. Some of these were in laboratories or in various stages of development. Therefore a "call for good ideas" went out while operators and developers collaborated to evolve operational concepts, doctrine, and required capabilities for the TF XXI.

Though activities on the TF XXI SOS architecture were underway, the cut-off date for initiatives to be incorporated did not occur until June 1995. The completion of developments or enhancements and testing of individual systems occurred between January and June 1996. By June 1996 these systems were scheduled to be in place for the final SOS integration at Fort Hood to support a subsequent series of exercises preparatory for the NTC event. The discussion of these events surrounding the integration will be given in the next chapter.

Task Force XXI Architecture

The process to define the operational, technical, and systems architecture for the TF XXI experiment was underway in January 1995. In one sense it is misleading to infer that all the necessary preparation occurred after this date. Rather the accomplishment built upon the lessons from the Louisiana Maneuvers, which began in Spring 1992, the previous events of the TF XXI plan,

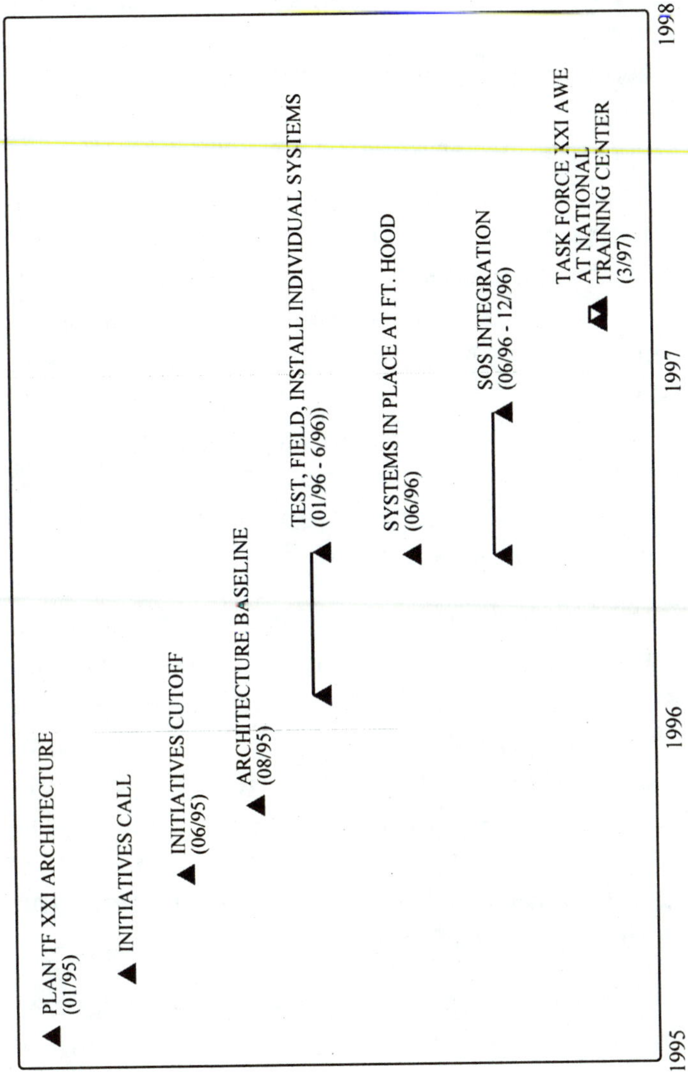

Figure 3-4. Task Force XXI Architecture Schedule

and the earlier efforts underway to establish a technical architecture for the Army.

The essence of the experiment had to be determined, and then translated into key digitization capabilities that were essential in the systems architecture to enable the experiment. Key to defining the operational architecture was the postulation of how a brigade would conduct operations with all the assistance of information technology. Operators had to determine what information was needed and how it would be used. Mission threads had to be examined end-to-end. Doctrine and tactics needed to change, too. The operational architecture for TF XXI focused on the nature and form of the information required, how operators would actually exercise their functions, and operational and organizational relationships. Performance bounds were assessed as well.

The implications of the evolving operational architecture and the core capabilities had to be analyzed and engineered with a view to determining needed changes to existing systems, new developments, and initiatives, while ensuring interoperability, capacity, and performance. A TF XXI system engineering master plan emerged. The substantial participation and coordination needed was achieved through various forums and through integrated product teams.[19]

The technical architecture used was the (then) defined Army technical architecture,[20] versions of which

[19] Examples of these included fires, chemical, aviation, mounted battle/armor, and communications/signal.

evolved over the duration of the experiment (Army Technical Architecture, 1996). At the time the TF XXI architectural planning began and while efforts were underway, the Joint Technical Architecture (JTA) for the defense enterprise had not yet been issued. However, the Army's technical architecture was used as a primary source[21] for the initial version of the JTA, mandated in August 1996. The TF XXI AWE was among the earliest of programs to field a large-scale systems architecture based on a defined technical architecture analogous to the JTA.

The Army's technical architecture[22] provided a minimum set of standards to facilitate integration of the systems of the TF XXI. Where possible, the architecture used commercial standards and provided a framework for information modeling and data exchange, for information processing, information transport, human-computer interface standards, and information security. As examples, the information processing standards addressed a distributed computer environment for UNIX systems. The information transport standards specified the Internet protocol. Data and message standardization was provided, including the definition of a variable message format and tactical

[20] Version 4.0 of the Army's technical architecture was published in January 1996, and Version 4.5 in November 1996, which was in use at the time of the NTC event. This subsequently has evolved to the Joint Technical Architecture–Army.

[21] About two-thirds of the joint technical architecture was derived from the Army's technical architecture.

[22] The scope of the Army's technical architecture was principally, but not exclusively, focused on C4I systems.

digital information link messages. A command and control core data model was developed. Key messages were defined, such as a "call for fire."

The technical architecture incorporated the DII/COE concept. The then-available version of the DII/COE was not used in the TF XXI; however a similar layered framework was implemented using many of the same software products. Others were surrogates, many of which were commercial products. The net result was that a common operating environment and common services were provided as part of the architectural framework for the SOS.

The architectural process was a spiraling one, continuing through the integration phase, and adjustments in operational and systems architectures were made until the conclusion of the SOS integration (Boutelle & Grasso, 1998). For example, the human-computer interfaces were enhanced with features tailored to the user and with common symbology. A switch-based network was migrated to a commercial router-based network for the operations centers. Solutions for firewalls and intrusion detection systems were completed after experimentation with multiple commercial products. System administration, directory services, and start-up were improved. And changes were made to tactics, techniques, and procedures.

The key revolutionary operational capability introduced for the AWE was situational awareness, achieved through the integration of a number of key systems and components into the overall architecture. From these all soldiers could derive a near-real time common picture of the battlefield—to know where friendly forces and enemy forces were positioned. The locations of friendly forces were automatically transmitted. The enemy locations were identified using intelligence, reconnaissance, and surveillance assets. Position location devices were linked to the Global Positioning System (GPS).

An integrated brigade command and control capability—part of the Army Battle Command System (ABCS)[23]—was provided for the brigade task force in a dozen Tactical Operations Centers (TOCs). These were connected to maneuver, air defense, artillery, combat support, and intelligence systems, and in turn, to the division command and control systems. This is shown in figure 3-5. The information was disseminated with a mobile router-based network, also illustrated.

The networks linked the TOCs to the level of the individual soldier and vehicle, and allowed command and control messages and information to flow. Messages flowed in various formats, and message content ranged from position reports to graphical overlays. In turn, automatic position reports flowed upwards through the battalion and brigade TOCs to provide the common picture of the battlefield at various levels.

In addition to ABCS, the core capabilities included:

---

[23] A critical aspect of digitization is to structure the Army Battle Command System to allow seamless connectivity across echelons and connect to the joint defense enterprise Global Command and Control System.

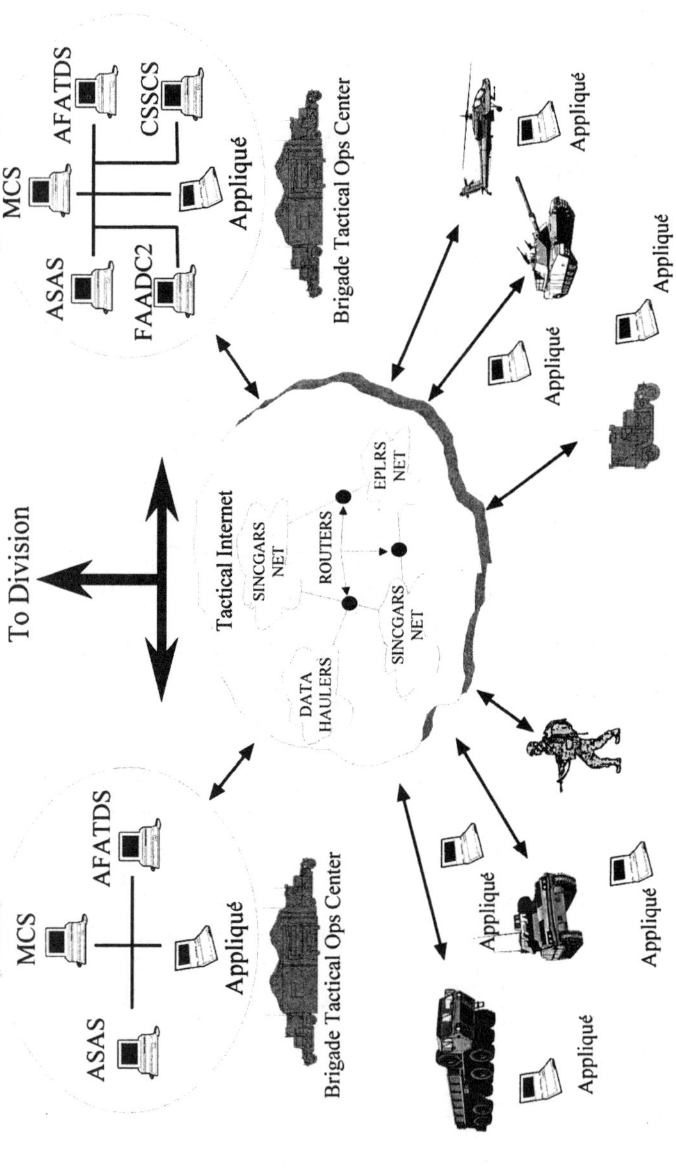

Figure 3-5. Task Force XXI Tactical Operations Centers

- Digital appliqués, which were small processors in four different versions, mounted on all vehicles or carried by dismounted soldiers, interfaced to GPS, and connected to a communications system of digital radios
- A Precision Lightweight GPS Rreceiver (PLGRS)
- An Enhanced Position Location Reporting System (EPLRS), which was jam-resistant and secure, used for data hauling and for providing near real-time position reporting of the tactical forces
- A Battlefield Combat Identification System (BCIS) to differentiate equipped friendly forces from foes
- A Single Channel Ground and Airborne Radio System (SINCGARS) Improvement Program with a commercially-based Internet controller for digital communications, and
- A Tactical Internet based on commercial routers and switches linking all these computers, radios, satellite terminals, and reconnaissance/ surveillance platforms.

Situational awareness was achieved for dismounted troops, for vehicles, as well as for various platforms through the integration of these core digitization capabilities, as illustrated generically on figure 3-6, which shows them incorporated into a weapons platform.

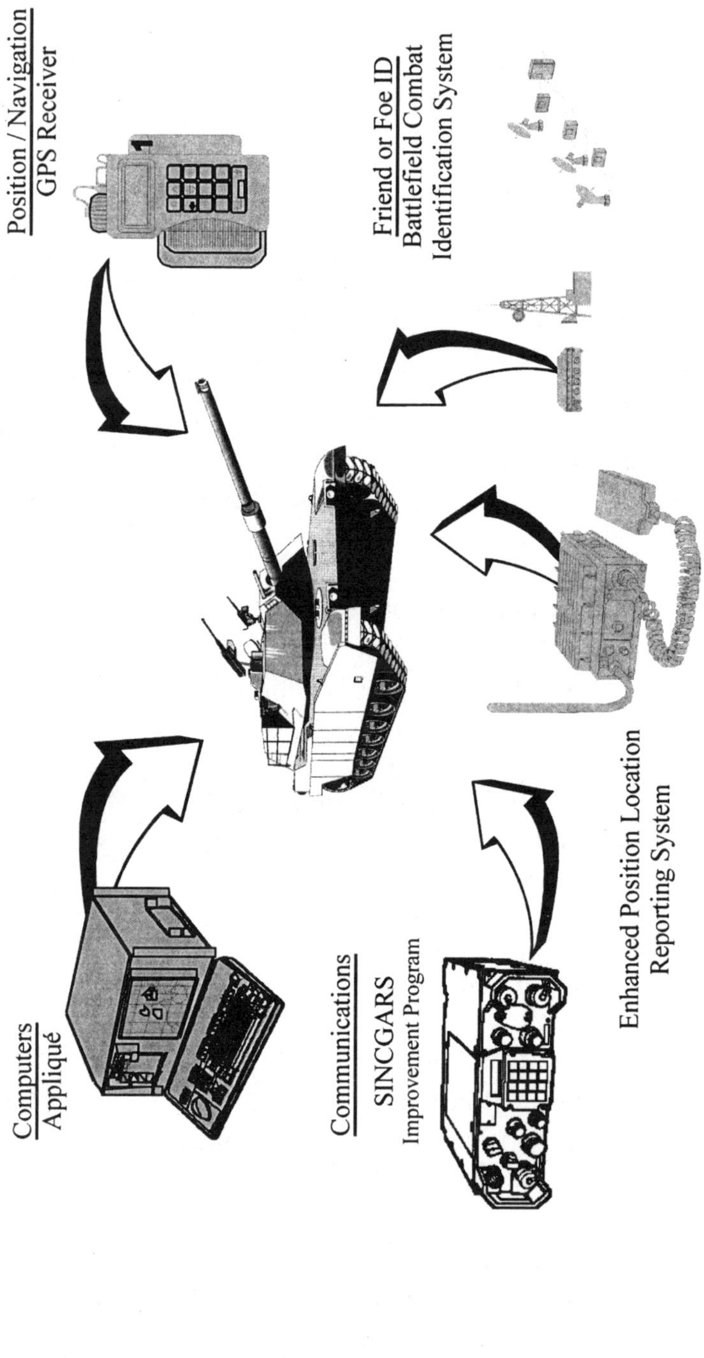

Figure 3-6. Task Force XXI Core Capabilities on a Weapons Platform

The TF XXI architecture fielded at the NTC was an aggregate of enhanced legacy systems and 72 separate digitization initiatives, many of which were individual systems that included developments of several years' duration. Some of these were concepts or systems that were tried in previous experiments but were transformed by the newer digitization technology. For example, in the late 1980s, the transmission of target locations had been accomplished by using a radio system linked to ground vehicles with computers and position location systems (Holcombe, 1998). This concept was significantly advanced through the core digitization capabilities of the TF XXI.

At one point there were as many as 171 initiatives that were part of the TF XXI architectural baseline, in response to the call for good ideas. These were gradually winnowed to 72 to ensure that the systems deployed to the NTC were sufficiently mature to support concept experimentation in the field. Screening experimental capabilities sufficiently to support their fielding was a lesson learned from the Desert Hammer and Warrior Focus experiments.

The initiatives were not only diverse in technology, but required multiple configurations, given the various platforms and vehicles. Two references describing many of these are Goedkoop (1997) and Hester (1996). Initiatives ranged from a mortar fire control system to a tactical end-to-end encryption device to a ground control station for the unmanned aerial vehicles. Of the 72, about 60 were innovative digital information systems. A sense of the breadth of the experiment is graphically communicated by figure 3-7, which is provided without further elaboration (TF XXI outbrief).

There were organizations external to the Army that also participated and/or brought experimental capabilities to the TF XXI complement of systems, although these were limited in number. These included elements of the U.S. Marine Corps, the Special Operations Force, the Defense Advanced Research Projects Agency, the Air National Guard, and the U.S. Air Force.

The scope of the systems architecture was vast, comprising hundreds of systems, including those providing fire support, air defense, maneuver, logistics, command and control, communications, and intelligence. There were more than 5,000 pieces of equipment with more than 900 vehicles and 180 different configurations required to support the activities at the NTC (Hanna 1997). Thousands of pieces of new equipment were installed on existing platforms[24] *"including 873 appliqué packages, 336 EPLRS, 1,550 SINCGARS, 62 BCIS, and 1,386 other types of TF XXI equipment"* (Goedkoop, 1997). The details of the architecture are described at the TF XXI web site.

[24] Platforms included wheeled and tracked vehicles, aircraft, and even personnel. Many existing platforms were used but had to be retrofitted with the key digitization capabilities. Some initiatives required new platforms.

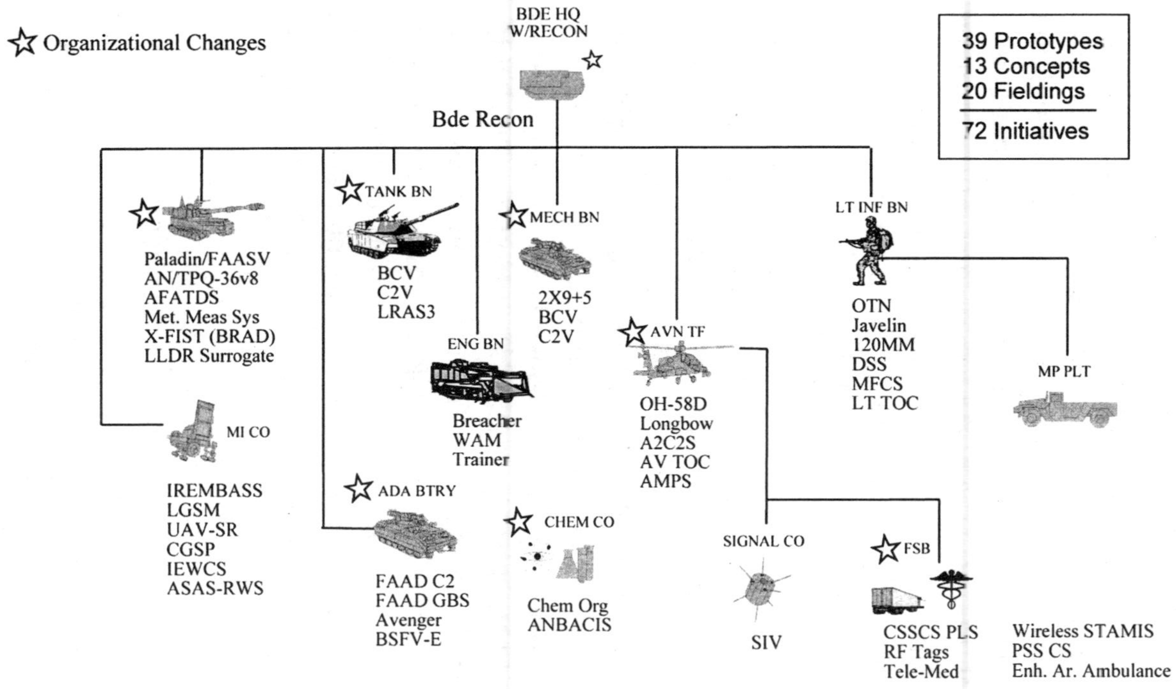

Figure 3-7. Task Force XXI

After the Experiment

The TF XXI AWE occurred at the NTC in March 1997 as planned. There were about a thousand official observers and controllers, augmented by subject matter experts. Substantial data collection enabled evaluation of the results afterward. The key question that had to be answered was: If information age battle command capabilities and connectivity exist across all battlefield operating system functions, will increases in lethality, survivability, and tempo be achieved?

Numerous additional questions were asked about the impacts of specific technologies and weapons as well as about the effects of force structure, doctrine, organization, and tactics, techniques, and procedures. To obtain answers, data were compiled not only from events at the NTC, but from predictive and post-NTC constructive simulations, as well as from assessments of EXFOR training events before and after. There were special reports generated, including one on training effectiveness. The analyses and final reports addressed the *potential* of digitization, relying not just on the actual force-on-force encounter at the NTC, but on the opportunities to replay some events later using simulations.

There were differing views on the degree of success of the experiment, not unusual for an undertaking of such magnitude. Some observers did not attribute an operational advantage to digitization, at least to the then-fielded version. The interested reader will find a rich stock of publications[25] available for perusal. The Army's executive summary of the results is provided in Hartzog (1997), which highlights the achievements without dismissing the problems and challenges. The immaturity of certain capabilities along with the connectivity problems did impact the activities at the NTC; however, the Army viewed the TF XXI as an experiment, not as an operational test. Its assessment, which used qualitative and quantitative data, acknowledged the tremendous potential of digitization, and on balance credited the overall success as much greater than any specific failure. The executive summary stated:

> *The TF XXI AWE was a highly successful experiment that exceeded the expectation of planners and participants alike. Not only did it reveal a clear vision of the dynamic potential in the digital land force, but it incidentally validated the Army's whole approach to experimentation.* (Hartzog, 1997)

Management Structure for the Task Force XXI

The TF XXI AWE was primarily an Army event, which simplified the lines of accountability. Many Army organizations participated, reflecting the priority accorded the battlefield digitization program. Strategic guidance and direction came directly from the Army's senior leadership, which comprised a forum akin to a

[25] Stanley (1998) provides some first-hand comments. A selective bibliography on TF XXI compiled from many sources is available on the Internet (Gibish, 1997).

"Board of Directors" chaired by the Army's Chief of Staff. The Army's Digitization Office (ADO) coordinated the integration of digitization activities across the Army.

The Commanding General, Training and Doctrine Command (TRADOC), had the overall responsibility for the TF XXI experiment. He was supported by the Forces Command (FORSCOM) and the Army Materiel Command (AMC). The Program Executive Officer for Command, Control, and Communications Systems (PEOC3S) was accountable for the systems architecture supporting the TF XXI AWE although other program executives provided essential capabilities such as for air defense and aviation. Because one PEO was designated accountable for the SOS architecture, the decision processes and integration of capabilities were simplified **behind the Wizard's curtain**.

The Director of Information Systems, Command, Control, Communications, and Computers (DISC4) evolved the technical architecture of standards and guidelines.

The convergence to the operational architecture and systems architectures for the experiment relied on teamwork. An experimental working group, with general officer participation, provided definition and direction of the TF XXI for TRADOC. A TF XXI process action team and a coordination cell planned the experiments, finalized the set of initiatives to be used, refined the mission threads with tactics, techniques, and procedures, and coordinated with the users. The systems

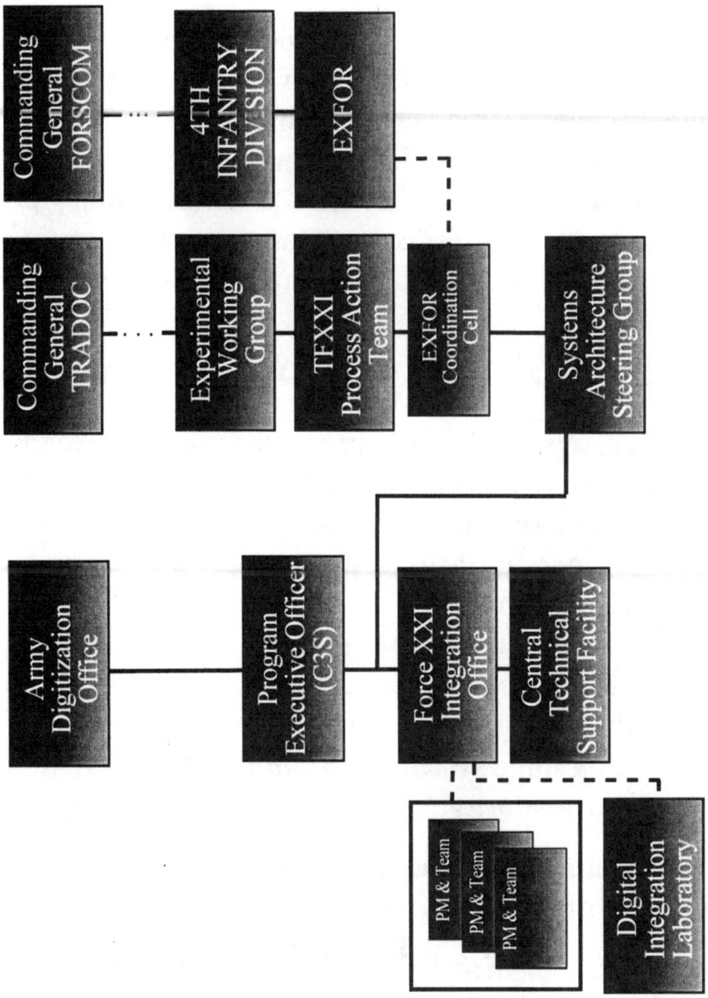

Figure 3-8. Task Force XXI Architecture Management

architecture steering group, managed by the PEOC3S, with participation from the ADO, the DISC4, and AMC, factored the effects of the initiatives, coordinating the development activities required for the TF XXI SOS.

The "trail boss" for integration reported directly to the PEOC3S, directing efforts at the CTSF. The program managers, their teams and systems, came from many organizations to accomplish the integration on-going there, as described earlier. The Digital Integration Laboratory was one of several organizations supporting testing. Figure 3-8 highlights many of these roles.

Comparison of the Two Case Studies

The overviews of the two programs illustrate many characteristics in common. Both were ventures of considerable scope and boldness of conception. Their respective organizations developed and applied operational concepts that were dramatic departures from those of then-current operations. To implement innovative and leading-edge capabilities, they both attempted to bridge technological gaps. This resulted in integrating some systems and components of varying levels of (im)maturity.

The as-built architectures were considerable in size—thousands of pieces of equipment were fielded for thousands of users. The DPS when installed required more than 380,000 square feet of facility space. The TF XXI required as many as 600 railroad cars to transport the necessary equipment and supplies from Fort Hood to the NTC at Fort Irwin.

The DPS comprised the integration of about 10 million lines of code; the TF XXI required about 40 million lines of code. For both programs there was a combination of developed software as well as commercial off-the-shelf (COTS) software.[26]

Both programs were risky undertakings from many viewpoints, including technology and size. While the Army's venture was greater in scope, both the DPS and the TF XXI, if assessed as a single information system, would qualify as highest risk on the scale of difficulty.[27] Complexity was amplified by the relationships between and among their individual systems. For example, the DPS developers characterized it as tightly coupled because the various constituent systems were highly dependent on one another to execute their own activities. For TF XXI battlefield digitization, the command and control strategy applied has been characterized as tightly coupled because of its need for near real-time

---

[26] DPS used COTS for information processing and some communications capabilities when commercial products could meet performance requirements. The digital photogrammetric and cartographic applications required development of software and hardware. Later many such components were transformed into commercial products.

The Army introduced many commercial products into TF XXI to facilitate the SOS integration. Examples included the distributed computing environment, the TOC backbone, and client/server applications to complement systems like the ABCS.

[27] Per a model using function points, described in chap. 1. DPS and TF XXI fall into the project category of 100,000 function points and 1 million function points respectively, which results in characterizing them as high risk based on the performance of other projects of comparable size.

synchronization of information (Czerwinski, 1996). The implication of a complex, tightly coupled structure is greater risk, unpredictable behavior, and vulnerability to failure according to Charles Perrow's classic quadrants' model (1984; Czerwinski, 1998).

Both the DPS and the TF XXI are characterized as a SOS. They were managed, developed, and operated by one organization, i.e., one agency and one service, respectively. Their development and operation were subject to direction, rather than collaborative in nature.

The DMA management structure was more centralized. As a smaller institution, DMA had only one internal organizational unit responsible for directing and developing the DPS. The Army used multiple organizations but focused leadership for the TF XXI by clearly designating accountability. Both organizations transferred considerable decision-making power to a few leaders to manage the respective ventures.

The DPS operational architecture was a postulation by DMA as to how the Agency's workforce would operate the SOS. The systems architecture included constituent systems managed, developed, and operated by the DMA. Yet each was a large-scale system in its own right, developed independently by different acquisition teams and contractors, under broad program guidance and direction. The independence of each system was amplified by freedom of design and development methods. The technical architecture, such as it was, imposed few standards and guidelines, a circumstance influenced by the technological challenges of the program. A common environment was applied to a small degree.

The operational architecture for the TF XXI was an Army view of a digitized brigade, although participants (and systems) external to the Army did engage in the experiment. The systems architecture encompassed many Army legacy and developmental systems, and many platforms and configurations. The individual systems served many different purposes, developed by various acquisition teams and contractors at different points in time, although under overall Army direction. The technical architecture was evolved by the Army and provided a set of standards primarily focused on C4I systems, but far more comprehensive than DMA's.

Figure 3-9 provides a notional comparison of the DPS and the TF XXI. The more centralized, almost unitary control and direction, moves the DPS closer to the origin on the autonomy axis. The TF XXI appears as more heterogeneous, with the breadth and numbers of different systems, and the use of legacy and some external systems. The DPS, developed using fewer than 10 sites and operated from three principal geographic locations, is depicted as less dispersed than the TF XXI. Although operated at the Fort Irwin NTC, the TF XXI SOS reflects operational scenarios with information assets geographically dispersed, even on a global scale.

There are differences in the two case studies relative to their overall objectives and methodology. The DPS was developed as a *production* capability. TF XXI was a *"product mature enough for an experiment to provide*

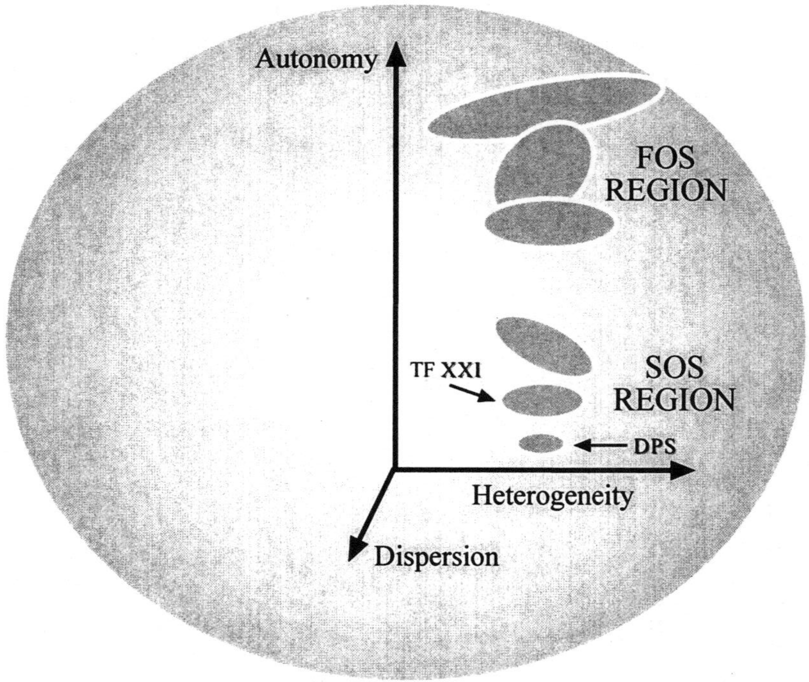

Figure 3-9. DPS and Task Force XXI Comparison (Notional)

data to the Army leadership for making investment decisions..." (Boutelle, 1996).

DPS, developed in the 1980s, generally resulted from a classic methodology characterized as a waterfall, and required nearly 10 years to deliver. The architecture developed over time from a bottom-up approach building upon the individual systems and their interfaces. The TF XXI architecture began top-down and resulted from a spiral evolutionary development process, which itself became a by-product of the experiment.

Chapter 4

The Integration Experiences

This chapter describes the events that transpired during the integrations of the DPS and the TF XXI. The narration builds on the overviews of the two programs, their architectures, and their organizations as described in the previous chapter.

With an understanding of the challenges both these programs faced, it is now possible to examine how each approached the integration of the individual systems in order to achieve a powerful new entity—a system of systems (SOS). This chapter explores the considerable efforts undertaken to plan and prepare for each integration and what actually occurred. Among the characteristics that both ventures shared is that neither had prototyped the integrated product previously and this circumstance made the integration task even more difficult.

A single integration facility with its supporting environment of people, processes, and infrastructure was essential to manage the integration successfully. In addition, in the TF XXI case, the environment fostered an evolutionary acquisition and development process. This chapter discusses these experiences.

Defense Mapping Agency's Digital Production System

Before Integration of the Digital Production System

The DPS behavior was anticipated to be the union of the behavior of the individual systems that comprised the SOS. The DPS was perceived by developers and users alike as a digital processing pipeline. The whole was expected to be the sum of the parts. Each individual system would contribute its required functionality within its allocated timelines and in a particular sequence. In simple systems where relationships are linear, this is more generally true (Czerwinski, 1998).

Between the time of the DPS critical design review and the start of its first integration event for initial operating capability (IOC), a series of formal requirements verifications and functional tests were accomplished for each individual system and its interfaces to other systems. This series was repeated before the full operational capability (FOC) integration as well.[1]

In addition, a complementary and independent series of activities occurred to verify the correctness of the interfaces. Because of the data dependencies among the

[1] At IOC, some interfaces, primarily those associated with the production management system, were not defined. Their testing and verification was completed before the FOC milestone.

individual systems, the exchange of simulated and real data provided an additional verification for the correctness[2] of the interfaces. One such activity required that the systems generating data provide that data to the systems consuming it, a check that consumers and providers interpreted data format and content identically on both sides of the interface. A second verification activity used an independent agent to interpret the interfaces and generate data in accordance with that interpretation, again requiring the consumers to verify and the producers to corroborate. A tool called a "scenario generator" was used for this testing, which was conducted at the various factories of individual system developers. Some of these verification activities transpired as much as a year before the first SOS integration phase. However, the early exchanges were hampered by late closure of some interfaces. Nonetheless, these efforts were critical in resolving hundreds of interface-related issues.

Also before the integration, a few subsets of the SOS were integrated and tested, as were common components shared between individual systems. Certain services common for all systems in the SOS were installed and tested at the various factory sites. The network services subsystem was one prominent example.

A single site was planned for SOS integration, a new Production Center, not yet operational. This reduced

---

[2] Verification of interfaces focused primarily on syntax; however, the development of a MC&G data model enabled verifying data messages of that type for content as well.

the impact of integration activities on the two existing Production Centers, where significant production was underway. The first delivery of the integrated SOS was intended for the new Center, where the almost exclusive focus was on transitioning the DPS to production use. The new Center was similarly used for the second integration event for FOC. The facility comprised about 200,000 square feet of space, half for the installed DPS, the other half to support integration activities. The size of the DPS facility requirement was indicative of the magnitude of the integration event.

At least 2 years before the SOS integration, detailed preparation for its verification was underway. A series of demonstration events was defined. As each event was successful, it signaled the SOS readiness to support the mission. Some demonstrations focused on formal verification of all the information exchanges, formats, and content; others focused on inter-Production Center relationships. Based on engineering analysis, operational jobs and tasks were defined for inclusion in key demonstrations used for both the IOC and FOC. These were constructed to test the functionality and requirements for the DPS as exercised by operators in end-to-end threads of operational activities. The schedules for each integration were estimated based on projections for each task in these demonstrations, factoring product difficulty, skill levels, and margin for reserve.

The SOS integration was anticipated to be difficult. This was well recognized at least 4 years[3] before the FOC. DMA management had wrestled with difficulties in the

Mark 85 program, a far less ambitious undertaking. The plethora of misinterpretations of MC&G feature data caused difficulties in the earlier program and provided strong incentives for early interface testing. The data exchange programs for the Mark 90 DPS, as well as early integration of some subsets, were used to reduce risk and resolve many problems before the integration, which they did.

DMA's program management expected that the communications services of DPS would be problematic, and that interoperability issues would arise from the diversity of platforms and operating systems. Because many individual systems had technological challenges, the constituent systems were at varying levels of maturity. Their residual defects were expected to contribute problems despite their assessed readiness.[4] All participants recognized that the SOS integration schedule was compressed and risky.

The Start of the Digital Production System Integration Event

The installations and tests of individual systems at the DPS integration site were time-consuming and resource-intensive. Their engineering teams moved to the site to

---

[3] An advisory board chartered by the Director of DMA assessed the DPS integration as very risky and recommended early integration efforts.

[4] A review ascertained that systems met requirements and were sufficiently robust to ship for installation and DPS integration. Discrepancies were assessed as acceptable or unacceptable before shipment Unacceptable discrepancies were corrected before shipment; nonetheless, many defects remained.

complete the integration testing with the other systems, and to support the SOS demonstrations. *Informal* testing among the individual systems began as soon as possible because of the difficulties anticipated.

When the systems were linked to provide initial operator access to the SOS, assign jobs, and initiate the flow of imagery, a staggering number of problems not previously manifested occurred, despite the earlier testing on individual systems. The integration event revealed the naivete of the assumption that the SOS would behave as the sum of the constituent systems, that the proper functionality would occur, and occur in the sequence anticipated. While difficulties were expected, the DPS behavior appeared unpredictable, and the nature and number of problems were confounding. Ascertaining the source of problems overwhelmed those reporting them.

The systems providing services were tightly coupled with those systems using them. The data in one system affected and altered the behavior of another. In many cases responses were unanticipated if the data were out of range or interpreted in different and unexpected ways. Varieties of reactions from anomalous data occurred that resulted in significant delays of hours, even days, in the flow of imagery and management data among the various systems, and even in permitting legitimate access to the systems. Because the DPS was intended to be a map-producing engine fueled largely by imagery, this initial situation was catastrophic.

Problems mounted while personnel threaded through logons and job assignments. Systems with workstations awaited scheduling, or when scheduled, awaited data, or when data arrived, prompted unexpected results,[5] not all of which were adverse. Dependencies imposed by the centralized production management and data and communications services were difficult to anticipate until the sequence of end-to-end threads of operations were exercised in the demonstrations used for integration.[6] Then it quickly became clear that the effects were definitely underestimated in severity.

The Integration Event Produced a System of Systems Prototype

The phenomenon observed at the integration site was akin to the birth of a new personality—the SOS—which subsumed or altered the behavior of the individual systems. It became clear that, as an entity in its own right, the SOS had received insufficient attention when compared to the attention directed to the individual systems. The importance of having a prototype of the

[5] Not all unexpected results were viewed as adverse. When cartographers saw the rich imagery detail at the new workstations, they extracted more information than planned; however, this lead to unintended consequences on production timelines and mass storage requirements.

[6] The production management system managed and scheduled thousands of production events that required resource deconfliction. Each event had dependencies with predecessor and successor events, all frequently changing. As the DPS behavior emerged, the numbers and complexities of these relationships increased, resulting in a combinatorial problem. As a result, a complete set of SOS end-to-end threads was virtually impossible to identify, maintain, and test.

entire SOS became obvious with all the prescience of hindsight.[7] In one sense, the IOC integration event produced the first prototype.

Did this happen because the individual systems of the DPS were not well understood? No. Their individual behavior was far better anticipated. Many systems benefited from previous development initiatives that had been successfully implemented in the earlier Mark 85 phase. Many had been prototyped with operators providing feedback to the developers at the factories. What was not prototyped nor well understood was the DPS SOS as a functioning entity.

An equivalent revelation occurred in the *second* SOS integration for FOC. After the IOC concluded, the developers added functionality to the individual systems. The production management system, in relative contrast to the others, had substantial changes[8] between IOC and FOC. In addition, the FOC milestone signaled the readiness of the new three Production Center operational capabilities (and new operational interdependencies). With such changes, the second integration event was as difficult as the first and the SOS behavior equally confounding; it was, after all, a different entity. Nonetheless, the participants were better

[7] The DPS Mark 87 prototype was cancelled when it could not be developed in time to support the individual systems manufacturing cycles. Integration revealed the importance of the prototype in providing insight into the behavior of the SOS.

[8] Because of schedule difficulties, only minimal system production capability was delivered for IOC.

prepared and the environment supporting them was considerably enhanced.

Operational and Engineering Leaders at the Integration Site

Of paramount importance to the SOS integration was authoritative leadership and a unified team *at the integration site*. The team had to be seamless between operational and development communities, and between the teams managing the individual systems and managing the SOS.

Before the onset of the SOS integration an Activation Control Team (ACT) was established to support the transition from development to production capability. The ACT was operating at the integration site nearly 2 years before IOC. The Chairman of the ACT functioned as a commanding officer to provide a focused voice for the operational community. A senior engineer, who reported directly to the Program Executive Officer, represented the development and acquisition community. Both leaders were positioned at the integration site to facilitate overall progress.

The problems resulting during the integration of the SOS demanded an almost continuous decision-making process *at the integration site*. It was essential to determine the causes of anomalies in functionality, develop solutions, and allocate responsibilities for actions to the teams of the individual systems. Requirements had to be reviewed to preserve the needs of the operational community as engineering decisions were made to accomplish the integration. The views of the operational community had to be clear because the individual system capabilities were changing, and the SOS was in a dynamic state.

On-site processes for decisions were coordinated with acquisition processes to maintain program control while supporting the resolution of engineering issues. Daily engineering boards, acquisition boards, and transition activity boards were all augmented with crisis-like schedules to accommodate the continuous coordination of testing and integration activities. Removing conflicts, setting priorities, and scheduling events required an iron hand.

A real threat was the prospect of a marching army of engineers and developers halted in their tracks by problems, with the effect of costly schedule delays. Authoritative decision processes, well-organized issue dispositions, and alternative plans for daily activities were among the means used to avoid unnecessary impacts and optimize time and resources.

At first the DPS integration was a confrontational process as the engineers of the individual systems, surrounded by users of many different persuasions, required resolution of issues, the source of which and the disposition of which were not immediately obvious. The importance of the on-site engineering board operating at a single geographic location cannot be overstated. The Senior Engineer was responsible for the SOS and had the authority to adjudicate between the competing demands of the individual system managers, who were given "face time" to present their analyses and recommendations.

The Need for An On-Site Engineering Board

The DPS Engineering Review Board, lead by the Senior Engineer and staffed by the SOS cadre and contractor assets, determined the responsibility for solutions to technical issues and problems in the SOS. This was a non-trivial undertaking because it required an understanding of the SOS entity, the means to determine the cause of the problems, and the means to understand how to resolve them. This staff was small[9] compared to those management and support personnel for the individual systems. However, it was a staff with knowledge of the DPS as a SOS.

In contrast, program managers and engineers of the individual systems had spent years managing their ventures, and identified strongly with them. They and their teams were not in the best position to assess the needs of the DPS, although all were committed to make it a success. Their diverse views of the SOS, like the tale of six men viewing the elephant, were comprehensive but specific to individual systems. They were less able, as a result of their specialized perspectives, to grasp the holistic behavior.

The individual system development teams also felt significant ownership of their own issue solutions; again this was a situation best resolved by the SOS

[9] One-twentieth of the personnel affiliated with the individual systems.

Engineering Review Board. It was important to maintain objectivity and understanding to arrive at the optimum solution for the SOS.

Any disputes (and the resulting funding consequences) of the SOS could be appealed to the higher authority of the DPS Program Executive Officer. This was invoked rarely because the on-site leadership was recognized as empowered.

Adding People to the System of Systems Cadre

For the DPS program management, full realization of the *level* of resources required and the *extent of expertise* needed to make a SOS out of a set of individual systems did not come until the beginning of the IOC integration activities. Each day brought more problems requiring disposition and resolution. More people were required to handle the increased activity. Confounding the participants was the circumstance that causes and effects of problems among the various systems could not be determined easily because of the complexity of the DPS.

Not only was there a need for *more* people, but for people with knowledge of the DPS as a whole (in contrast to that of the individual systems). They needed to understand the end-to-end behavior of the SOS that resulted from interactions among individual systems. Unexpectedly, a partitioning of knowledge about the SOS had occurred even in the SOS cadre. Members had worked so long with the teams of the individual systems that their own knowledge had become specialized.

The amount of information an individual needed to absorb and understand overwhelmed many. By the time of the integration, many individuals had participated in the program for nearly 10 years. They had engaged in DPS reviews, technical exchanges, interface implementations, and pre-demonstration activities. Yet despite this, they were at a loss to explain behavior that went well beyond the constituent systems.

The on-site operational leader characterized the situation as follows:

> *We had some idea of how many (people) were needed and did in fact program for them, but what was unknown was the amount of information people had to absorb and understand.*

Attempts were made to supplement the SOS cadre, but at the time of greatest demand, knowledgeable talent was in short supply. Understanding the SOS in all its manifestations required vast amounts of time, and this detailed knowledge was exactly what was needed for the integration. This shortfall was unanticipated and therefore not addressed, even in an elementary training program or in one that would have transferred expanding knowledge. With schedule pressures and the number of concurrent activities at integration, this was problematic. The situation resulted in performance above and beyond the call of duty for those who were in a position to contribute, and dedicated people were consumed by the effort.

Problems Understood and Resolved Faster Using One Site

The physical site for integration with its supporting environment afforded the opportunity to learn the emerging SOS from first-hand observation. It offered immediate access to details of behavior, allowing more rapid resolution of problems to prepare the DPS for production readiness. The demonstration leader, reflecting on the experience, said:

> *Problem resolution required continuous face-to-face communication.*

The need for communication, coordination, and collaboration among participants was intense. Information exchange and analyses of issues were best accomplished with on-site personnel who were in a position to observe, understand, and resolve. This was a shift in strategy because previously key engineering resources remained at factories where the robust infrastructure and large resources were positioned to respond.

Participants were prepared to deal with the complexities attendant to their individual systems. This was understandable because these developments were substantial undertakings in their own right. Each required expertise to develop; each was technologically challenging; each had an individual user community to satisfy; each had specific functionality to demonstrate; and each had performance and reputations tied to the success of the individual systems.

However, the SOS integration event demanded that engineers pool their diverse and detailed knowledge of individual systems to begin to achieve understanding of the integrated product—akin to dynamically developing a knowledge base about the DPS. The environment at the one physical site enabled the engineers of individual systems and the SOS cadre to compile, exchange, develop, and aggregate the diverse insights to describe the whole. Then, in turn, the on-site teams for the individual systems translated these understandings for their counterparts at the factories of the individual systems.

One Site Helped Make a System of Systems Team

The environment was essential to reforming a new team—the SOS team. The location of the integration at one site coalesced team work. It helped shift participants from their more parochial individual system views to one focused on the SOS. It helped transform many teams dedicated to succeeding with their parts to one team intent on succeeding with the whole.

The synergy of allegiance to the national mission, total professionalism, and pride of being part of one of the most challenging ventures of its kind emerged for all participants. Commitment, integrity, and self-sacrifice abounded. Today, even with the lapse of years, many participants still view it as one of the greatest and most rewarding experiences of their professional lives.

Management Tools at the Integration Facility

At the outset of the DPS program, a number of processes and practices were required by the program leadership to manage the individual system acquisitions. However, how these functions were implemented was at the discretion of individual system managers who had considerable latitude in design and implementation. For example, different tools and processes were used to accomplish project management, resource scheduling, action item identification and tracking, issue resolution, and configuration management. Many of these tools were brought in and installed at the integration facility by individual team managers.

The means used often reflected individual corporate practices of the developers. Because personnel affiliated with a particular system were familiar with and trained on specific tool capabilities in their own companies, they naturally selected these tools to support management of their acquisitions. Government teams managing these acquisitions then also became familiar with these individual implementations.

Limited Interoperability in Collaborative Mechanisms

At the outset of integration, project managers of the individual systems anticipated the need for their own supporting infrastructure at the DPS integration site. Their first priority was the relationship between their site team and their counterparts at their factory. The site team required key engineering resources and development assets to respond to the problems identified at the integration site. They planned and positioned themselves accordingly.

During integration, the priority quickly shifted to their relationships with the teams of other individual systems and the SOS cadre—driven by the needs of integration. Before turning to the factories to correct the problems, they needed more understanding of the nature of the problem, what to fix, and an agreed-to-plan. All of this required intense collaboration, communication, and coordination with their counterparts at the integration site through processes managed by the Engineering Review Board.

While their support processes and tool sets were appropriate for managing their own teams, they were not necessarily compatible for exchanging information with managers of other teams and the SOS cadre. This situation interfered with and slowed coordination and collaboration among the various individual system teams. The diversity in approaches subsequently was mitigated by decreed standards for selected information exchange as well as for common tool sets and systems specifically to support the needs of SOS integration and issue resolution.

While this facilitated the necessary collaboration, the shift to commonality resulted in the need to assimilate new tools and processes at a time (integration) when the resources and schedules of the individual teams already were stressed.[10] In addition to elevating schedule risk, cost impacts (acquisition of tool sets and personnel training) resulted. Difficulties also arose from reconciliation of the level of detail necessary to schedule activities and resources at individual system levels with those necessary at the SOS level. These impacts could have been reduced if the collaboration needs had been accommodated early on at the program outset and well before the start of the integration.

Examples

A good example is that of discrepancy tracking. The DPS program guidelines defined hardware and software discrepancies as defects in meeting program requirements. The methods, priorities, and timelines for their resolution were based on the level of severity of the defect. However the specific implementations (processes, tools, and systems used) to accomplish discrepancy handling at the factories where software and hardware were developed was left to the discretion of the management of the individual system teams.

For the SOS integration, it was agreed to migrate all discrepancy information to one computer system to control the entire SOS status, including that of the individual systems. The implementation of this migration lagged. As problems were identified during the integration event, what quickly became apparent was that the various individual system implementations

[10] While this need for commonality was foreseen before integration, the migration to common tools sets and common practices fell behind schedule.

had spawned differing interpretations of severity levels, categorizations, and the content of information required—despite exchanges among the individual system teams before the integration event.

When discrepancies arose in one system that affected others, collaboration and coordination were critical for rapid resolution and disposition. However, information and time were lost in translating the local dialects among the teams, then re-translating for their factory counterparts. These inconsistencies eventually were reconciled, but it required a year before all the information flowed smoothly among the various teams at the site and the factories. This situation burdened an already-stressed Engineering Review Board.

Another example is illustrated by the varieties of tools and processes supporting configuration management of the hardware and software baselines. The experience was analogous to that of discrepancy identification and tracking in that the variations in the individual implementations caused difficulties in managing the SOS baselines as a whole. There also was a need to differentiate the level of configuration item for which the management at the integration site would identify and control the baselines, versus the level required by the factory sites.

Adding Tools and Infrastructure to the Integration Facility

Several processes and automated tools proved beneficial in supporting integration management for the DPS. In addition to discrepancy tracking and configuration management, these included tools for requirements traceability, action item tracking, and scheduling daily events and resources (people, equipment, and facility space). The resource scheduling required the capability to project multiple contingencies for every event because activities did not progress as predicted. Eventually the SOS team established one configuration management system for the SOS baselines at the integration site, one SOS discrepancy tracking system, and one action item and issue tracking system.

Integration activities demanded a common means for communicating and documenting information. A common work environment unified the teams of many individual (and disparate) management infrastructures. Because the integration site was a Production Center, its office automation environment provided a common means for interpersonal communications for participants.

A telecommunications infrastructure united the Production Center that functioned as the integration site with the other two Production Centers. This was beneficial to link the Agency's managers, who were resident at all three locations. It was also essential to stage DPS software deliveries to all three sites. Later the infrastructure supported the Agency's operations when DPS was ready for production. An equally necessary communications infrastructure linked the integration site with the development factories of individual systems. This essential infrastructure supported interpersonal communications, video teleconferencing, and transmission of software changes, data, training materials, and documentation.

A Digital Production System "Core"

The implementation of a SOS "core" to augment the testing strategy was started before the integration event. This was a mini-DPS called the "DPS System Test Mode." Analogous implementations have been used on other information technology projects where there is a need to assess changes without perturbing the operational mission. The offsite facility in support of the Cheyenne Mountain Complex is similar in concept.

The DPS System Test Mode was a minimum set[11] of all the unique hardware components and all the software baselines of the individual DPS constituent systems. As such, it was a specifically configured standalone SOS comprised of all individual systems, detailed to their hardware and software components. It was equivalent in functionality to the operational DPS except for those configurations when unique hardware components were

[11] There were some caveats to "all the unique components." The major exception was the imagery fetch and dissemination capabilities that were not included in the system test mode as a dedicated asset, but could be redirected from production assets when needed.

deliberately omitted because they were unavailable. It had logically independent, parallel networks and physically separate data bases and could operate concurrent with, but isolated from, production use or other integration activities.

The System Test Mode supported end-to-end multiple thread testing for the DPS and included tools that allowed software flows to be altered, bypassed, or modified to facilitate testing of a particular configuration item. It was used for testing and build verification of the SOS and came under the management responsibility of the Engineering Review Board. While it was not available in time for the IOC integration, it was used for the second integration event before the FOC milestone, and subsequently for DPS maintenance and evolution. It was installed at the integration site, and later at the other two Production Centers, and was used for geographically distributed (inter-Center) testing. It was valued as a necessary tool of the SOS integration environment.

The purpose of the System Test Mode was grounded in verification, not for concept definition or design evolution, as a SOS prototype might be applied. It could be tailored into specific configurations as needed for testing except for certain kinds of stress loading and resource contention (because the equipment suite used was intended to be minimal). Consequently, it provided a capability to assess the effects of changes to the individual systems and to evaluate them on the SOS holistically using a complement of known results. When sufficiency of testing and correctness were established, the changes then were migrated to the installed SOS baseline.

As test results were gradually compiled, a sizeable set of benchmarks accrued. These eventually included all end-to-end flows and associated test data as well as training data. Engineering analysis was used to select the subsets of the benchmarks to be used, based on where the risk was highest, or the change greatest, or even which test was on the critical path. Consequently, the quality of the change processes improved dramatically and adverse side effects from changes could be determined earlier and with increasing predictability.

A key adjunct to the System Test Mode was its companion set of tools, including the scenario generator, which could initiate message traffic and file transfers, act as a passive or active receiver of service requests, and then respond accordingly.[12] This tool was evolved based on information gained from the earlier independent data exchange program for testing the interfaces between individual systems.

The System Test Mode proved invaluable to support evolution and maintenance of the DPS after FOC. After the production equipment was operational at the Production Centers, the hardware components automatically could be extracted from production use,

[12] It acted in accordance with the interface specifications. If a message received was intended to produce a certain response, that response was generated to the appropriate systems as part of the verification.

configured, and dedicated as testing and verification assets to assess the consequences of changes across the SOS. This avoided inadvertently introducing adverse side effects on the production floor.

Conclusion of the Integration Phase

The successful completion of a series of demonstrations comprised the conclusion of the integration phases for IOC and FOC. In addition to the message exchanges, the series included the production of four MC&G products for IOC and eight products for FOC. The products selected exercised major aggregates of functionality, were high priority, and were considered difficult to generate.[13] For example, a coastal chart was selected for FOC because it included aspects of both land and water feature attribution. In the pre-DPS era at DMA, some of these products required 2 years to produce.

The two demonstrations for IOC required 7 months; the four demonstrations for FOC required 10 months. Their completion, coupled with the correction of the more severe discrepancies, signaled readiness for the next program phase, which after FOC was production.

---

[13] For IOC, the four products produced were the 1:50,000 topographic linemap, digital terrain elevation data at level 1 resolution, and two types of point target graphics. For FOC, the eight products included a 1:50,000 topographic line map, a joint operations graphic-ground, a coastal chart, a digital feature analysis data product at level 2 resolution, a digital terrain elevation product at level 1, and three terrain contour matching products.

The U.S. Army's Task Force XXI

Before Integration of the Task Force XXI System of Systems

The difficulties of SOS integration were demonstrated again when the many individual systems intended for the TF XXI experiment first were integrated in June 1996. Systems arrived at the Central Technical Support Facility (CTSF) at Fort Hood for the final integration phase. Despite considerable preliminary testing, when the systems were interfaced, they failed in the first exercise, LZ Phantom, to support operators. One manager described the initial situation as *"disastrous!"* This circumstance stems from the same causes experienced in the DPS integration. This can be attributed to the complexity of the interactions of the information systems independently and individually developed and tested, and the emergence of behavior difficult to predict and comprehend when the systems are first integrated.

The Army had prepared for this event through an extensive and complementary series of tests and verifications of the participating systems, coupled with multiple training events. As described earlier, previous activities included a series of large-scale experiments beginning with Desert Hammer in 1994. Using a combination of Advanced Warfighting Experiments (AWEs), simulations (virtual, live, and constructive), and actual exercises, individual systems and subsets of systems gradually were tested. By advancing the digitization concept through exercises and experiments, the Army developed and integrated some systems required for the TF XXI SOS, testing them in operational settings. Holcombe (1998) provides a good reference for certain aspects of this progression. During the Desert Hammer AWE at the NTC in 1994, experimentation with a partially digitized armor force used: an appliqué-like capability; Tactical Operations Centers (TOCs) equipped with all-source intelligence workstations; and a Brigade and Below Command and Control System. Similar capabilities were later used in TF XXI.

In August 1995 Exercise Focus Dispatch centered on heavy forces and used a combination of virtual and live simulations with a field exercise. The digital passage of information was explored, such as for moving large amounts of imagery and graphics. A key objective was that of evolving techniques, tactics, and procedures, but this was complicated by connectivity and contention difficulties. One of the important by-products was progress on testing the communications system with that of the radio. This was one of the core systems included in the SOS for the TF XXI. The exercise exposed the problems of factoring "real world" effects, such as those from hilly terrain, into the communications traffic between tanks.

The Warrior Focus AWE at Fort Polk in November 1995 examined digitization in a light infantry environment. The results were used to improve command and control systems. During the experiment, a computer system linked to the GPS, the Brigade and Below Command

and Control System, a commercial communications technology test bed, and voice and data radio systems all were exercised. Also included were a lightweight Tactical Operations Center, versions of the Maneuver Control System, and a secure packet radio system.

The tactical internet, also one of the core capabilities of the TF XXI, was exercised at multiple events, including some using a variable message format. There was one virtual exercise of the tactical internet nodes for architectural compliance. All of these tests were used to identify and correct a variety of problems.[14]

In the early planning phases, a TF XXI System Integration Plan was developed that provided guidance for testing and experimentation to integrate the critical elements and systems (TF XXI Plan, 1996). It focused on risks and issues that had to be resolved for the SOS integration to proceed. Agreed-to methods to test resolution, such as analysis or modeling and simulation, were identified and used by the teams managing the individual systems. As the target battlefield architecture evolved for the digitized experiment at the NTC, each individual system in the SOS complement for the TF XXI also evolved by subsequent enhancements and by incorporating applicable results from previous tests and exercises.

Each initiative or modified system in the TF XXI SOS had its own integration, testing, and verification tasks

---

[14] The tactical internet did not achieve the level of reliability desired by the TF XXI event during the force-on-force encounter at the NTC. (Hartzog, 1997; Caldwell, 1997)

to complete successfully, and then each underwent another level of testing in preparation for an integrated SOS. Examples of testing for the SOS included that of the all-important message formats and data protocols, where considerable effort was expended. The integration plan was followed to resolve issues by the methods agreed.

The plan partitioned the SOS integration into logical components and subsets, such as the Army Battle Command System, the various platforms, the Tactical Operations Centers, and the tactical internet. Those initiatives not integrated through this strategy were individually integrated into the tactical internet. Figure 4–1[15] illustrates this approach.

Because the TF XXI focused on experimentation, there were many initiatives in the SOS complement that were in early stages of development. A set of success criteria was developed to ensure that these initiatives were sufficiently mature and deployable in time for the NTC field activities. Fifteen criteria were used in assessing factors such as safety, manpower availability, and readiness. As a result, the number of initiatives was pared to the final 72 in time for fielding at the NTC.

A Certification Process Also Established

When testing of the individual systems was completed, another stage of testing began. This was administered independently by the Army's Digital Integration

[15] Extracted directly from the TF XXI System Integration Plan (1996).

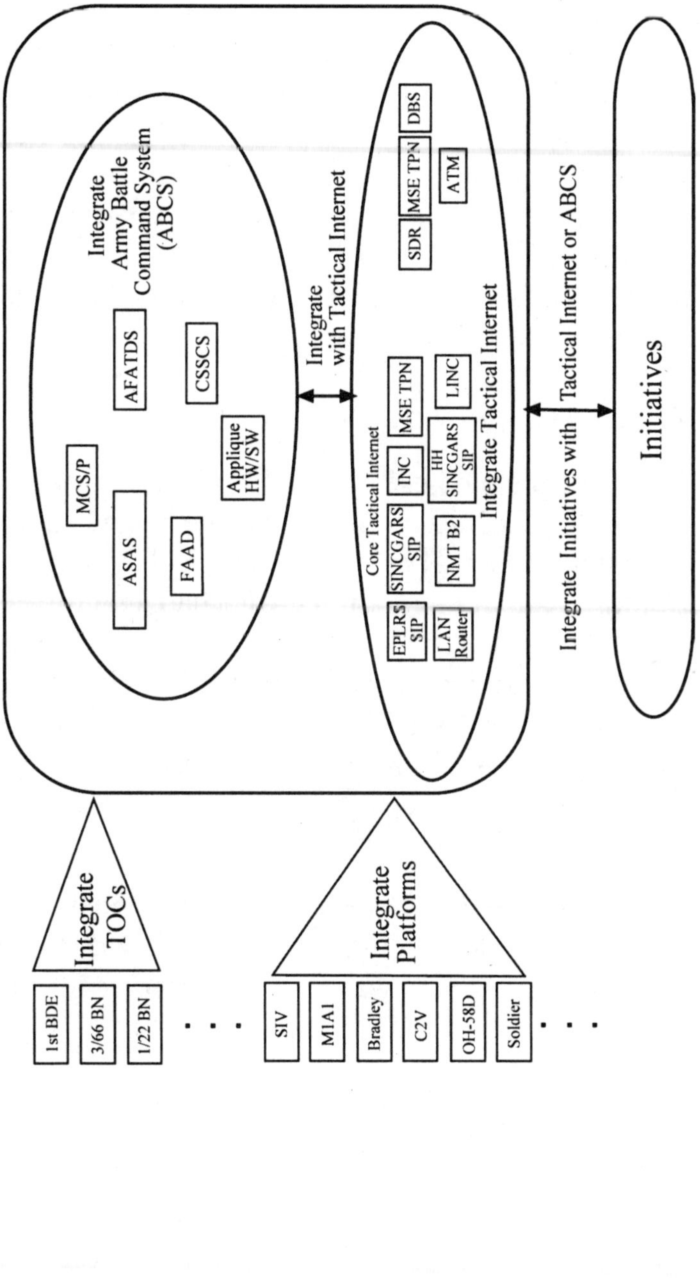

Figure 4-1. Integration of the Task Force XXI System Architecture

Laboratory (DIL) and generally occurred just before a system's shipment to Fort Hood.[16] This series of tests addressed compliance with the technical architecture, with military specifications and standards (such as the message format), and testing for interoperability. This provided an independent verification for each system and its interfaces. The DIL complement of testing was intended not only to ensure compliance with standards and architecture, but also to establish a certification process. Individual systems in the SOS for TF XXI had to pass DIL testing before proceeding into the final integration phase at Fort Hood.

Use of Fort Hood in Task Force XXI

Originally the planned use of Fort Hood was primarily for training for TF XXI, although the site had a history of integration support. The Army's Tactical Command and Control System Integration Office had a facility at that site. Previously it had supported large contingents of soldiers for training on new capabilities. However, the role of Fort Hood's CTSF initially was viewed as a nominal one for the TF XXI integration. It was to be used for correction of discrepancies in the software identified by soldiers in training.

Preparations for the TF XXI experiment previously had involved use of other facilities. Fort Hood was to be the place where the SOS was brought together for the first time in June 1996. All the existing systems would

---

[16] Some testing occurred before integration at the Fort Hood central technical support facility, which functioned as an extension of the DIL.

be integrated with the new capabilities to provide a SOS for training purposes as well as to evolve the final configuration for deployment to the NTC. The plan then concluded with a series of training events on the integrated SOS from August through December 1996.

The Reality of System of Systems Integration

In June 1996, the date on the overall schedule for everything to be in place, most systems[17] arrived at Fort Hood but few were ready for a SOS integration. This was demonstrated during the first exercise with the operational community. The LZ Phantom exercise conducted there in June 1996 provoked the comment, *"disastrous!"*

When the individual systems were hooked together, the SOS did not function. Despite all the previous testing, the TF XXI SOS could not support the Army's EXFOR in executing the operational mission. The complexity of the integration was fully revealed. The Tactical Operational Centers did not work well together. There was questionable performance between vehicles. There were difficulties in scaling to larger numbers of operational components, as well in supporting variable configurations, which numbered as many as 180.

The integration plan reflected the expectations that, given the circumstances of the earlier complementary suite of testing of the individual systems, the integration of the SOS would occur readily in a relatively short

[17] A major exception was allowed for the tactical internet, which arrived some weeks later.

window of a few months. The early days of the integration proved otherwise. All of these capabilities had been developed independently by program managers who believed they were building for integration. Despite their best efforts at developing or enhancing systems to fit within an overall architectural framework, testing systems and interfaces for compliance, and building upon previous exercises, the individual systems did not interface together, nor did they interoperate, fundamental requirements for a SOS.

Vitalizing Fort Hood's Central Technical Support Facility

When the reality of the challenges emerged, the solution was also at hand. The Program Executive Officer (PEOC3S) for the TF XXI experiment immediately responded to the problems by increasing the leadership focus on the integration, augmenting the engineering support, and evolving the role of the Central Technical Support Facility (CTSF) at Fort Hood. It now became a facility:

> *for rapid integration of dissimilar software and hardware systems, through real time interaction with soldiers, contractors, testers, program managers, and the requirements community.* (Boutelle &Grasso, 1998)

This decision increased the discipline and rigor of the integration process. In retrospect, this was a stroke of genius because it created the environment **behind the Wizard's curtain** to support the integration of the SOS, and thus its creation and delivery to the NTC in time to support the TF XXI exercise.

The PEOC3S also designated a leader in July 1996 to make the integration happen at the then-growing CTSF. This individual became the decision-maker "trail boss," adjudicating solutions to engineering issues, and moderating acquisition decisions to align with successful integration. This appointment acknowledged the need to arbitrate among the competing interests of the individual program managers and to prioritize the engineering of the SOS over that of the individual systems.

This empowerment was endorsed by the Army's Chief of Staff and the Commanding General of TRADOC, and it brought quick acknowledgement throughout the Army organizations, including all the participants **behind the Wizard's curtain**. These included the program managers of the individual systems, who controlled the bulk of the necessary resources with their contractor teams.

The response to the integration difficulties also included enhanced engineering resources. In fact it was a substantial response, sequestering many engineers at the CTSF and increasing the core staff to nearly 100 engineers. While approximately 1,200 personnel and 48 separate companies supported TF XXI overall, only several hundred personnel were resident at the integration site at any one time. Many remained at the development factories for the individual systems, or at other sites, tethered by communications infrastructure to the CTSF. However, there was a substantial enhancement at the site, and the CTSF grew markedly.

The TF XXI participants initially were apprehensive at the convergence of large numbers of program managers, engineers, testers, and developers at Fort Hood. However, one leader, on observing the team spirit that evolved there, commented on the magic of it—that some transformation occurred—and that everyone became dedicated to "making it happen." When one participant was asked how many teams were at the CTSF, he replied *"one."*

At Fort Hood the entire focus of the resident team became the successful achievement of TF XXI. The use of the CTSF with its supporting infrastructure generally is recognized as a major contributor to task force success.

Experimentation and Training at the Central Technical Support Facility

A tremendous synergy arose at the CTSF with the physical collocation of the operational community and the development community. The EXFOR in residence was partitioned into operational enclaves (such as battalion Tactical Operational Centers) for training. Soldiers engaged in a series of exercises that threaded operational activities end-to-end. The participants dubbed them "battle drills" because they provided substantive training experiences. In addition, they served to test the integration of the SOS.

As operators trained with new capabilities in mission-like activities on a daily basis, they provided instant feedback to the developers. Fort Hood changed into a factory-like setting, where vehicles and platforms were newly retrofitted, and the next day soldiers were trained on them. Developers stood next to operators. When capabilities did not work, they were modified by the developers, sometimes at the site, then reintegrated, and retested in an iterative spiraling process. Where necessary, additional functions were added. Operational and developmental compromises about the SOS sometimes were made in near real-time.

Many systems relied on commercially available components. Many pieces of equipment tested were commercial off-the-shelf, some were rugged versions and some were of military specification. But there was great diversity in vehicles and platforms, and accommodations had to bc madc to mount and install capabilities. Alternatives for installations were evaluated by soldiers to minimize interference when operated. There were more than 40 different vehicle types with new equipment and nearly 1,000 vehicles with about 180 different configurations (Goedkoop, 1997).

Operators were positioned to learn through experimentation and to use their knowledge to drive development. The developers gained considerable understanding of operational needs and responded. Both communities derived synergy through their collocation at the CTSF and the activities occurring there.

While this process was on-going, problems attendant to the integration continued, but resolution was focused by the leadership and this phase progressed. Activities in support of integration and training operated 7 days a week, 24 hours a day. The battle drills revealed defects in the integrated product, but gradually progress was made, which resulted in a SOS entity that was able to support more comprehensive activities. The events at Fort Hood concluded successfully with the brigade-on-brigade force-on-force exercise in December 1996.

On-Site Engineering Board Essential

The on-site engineering board processes, managed by the trail boss, were critical to the quality of architectural changes, resolution of technical issues, and acquisition decisions. Adjudication for resolution of issues and discrepancies occurred quickly and continuously. Having access to the right people at the site to resolve problems was essential. The requirements for engineering expertise were demanding. To fit capabilities within the overall framework where standards did not exist were anomalous, or were insufficiently detailed, decisions were made favoring emerging standards and future SOS integrations. The adjudication processes were open for participation. Coordination was facilitated by the presence of program managers and participants at the site.

The number of ongoing daily events necessitated a firm discipline for defects correction. As changes were propagated continuously, there was a need for rigorous configuration control of the SOS hardware and software baselines being integrated and tested. Over time, control increased in extent but was never completely managed by the CTSF because the change processes for some individual systems continued to be managed by their respective teams at their own sites or factories.

Exercising Authority Behind the Curtain

As discussed earlier, there was one primary Program Executive Officer for the TF XXI architecture. However, responsibility, management, and funding for the individual systems remained with the program managers of those individual systems in the SOS complement. Many of these managers, in turn, reported to other program executives such as those responsible for ground combat and aviation. The trail boss was the designated decision-maker for the integration at the CTSF, but the program managers retained the majority of the assets used. This resulted in a form of creative tension that required consensus for some decisions.

Because there were many systems in the TF XXI SOS, there were many managers and many different interpretations of needs and priorities. Some who managed legacy systems felt constrained because they had only limited funding programmed for a venture of such scope. This situation heightened concerns.

During integration, funding pressures increased as decisions resulted in unanticipated and unprogrammed changes to individual systems and interfaces. Requirements dynamically emerged or evolved. Sometimes the missions and funding for individual systems competed with the needs of the TF XXI SOS integration. Three factors eased this situation. First, the empowerment of the trail boss established a decision authority at the site. Second, the collocation of the operators and the developers improved the understanding of operational needs and reduced friction in favor of collaboration. Third, the participation of program managers and teams at the CTSF aided the resolution of disparate priorities and lent reconciliation to the decision-making process. The views and insights of the individual program managers were important to the quality of the decisions. Their availability supported the needs of a process that ran continuously.

Lines of authority and decision processes were well established and important to overall integration success. As discussed in the program overview, strategic direction came from the Army's most senior leadership, using a forum analogous to a Board of Directors, chaired by the Army's Chief of Staff. The "Board" provided authority and remained engaged throughout the TF XXI. An Experimental Working Group of general officers provided oversight for TRADOC.

An experimental coordination cell was comprised of a "Council of Colonels" to focus attention on key issues and to coordinate to achieve unanimity in the users' viewpoint. They wrestled with overarching TF XXI issues such as the effects of force restructuring and adjustments in overall schedules. They had recourse to the oversight boards as necessary. The Chairman of the Council of Colonels, who was resident at Fort Hood, was cognizant of the on-site operational issues and able to coordinate the operational viewpoint. The TF XXI Integration Office and the trail boss reported directly to the TF XXI Program Executive Officer.

Many Army organizations were involved with the TF XXI, and they were represented in the working groups, steering groups, and coordination cells that were established. The lines of authority were clear to the participants. The day-to-day decision making was made possible by empowerment and by teamwork at the site, and there were clear avenues to follow when issues appeared too contentious or significant. Generally this structure was very effective.

Integration Environment

The role of the CTSF changed dramatically in response to the challenges of the TF XXI integration. Because the schedule was compressed, management was not prepared to acquire an environment specifically adapted for integration, although the infrastructure was enhanced during the integration activities. This situation improved *after* TF XXI when the Army was able to assess the processes conducted at the CTSF as critical to the overall success of the experiment.

Certain infrastructure that contributed to management of the integration was in place early or evolved. One critical capability was that which supported configuration management of the changing SOS baseline. Another was the communications infrastructure. The latter tethered the factories of the individual system teams, which were geographically dispersed, to the integration site. It also linked certain engineering simulators necessary for integration, and connected to simulation centers that were used for training scenarios.

Initially the bandwidth of this communications infrastructure was severely limited but eventually became sufficiently robust to support video teleconferencing for the management teams. The CTSF was deliberately linked to the DIL, which in turn provided connectivity to many other sites.

Certain office automation support also was available. The CTSF supported participants with whiteboards, personal computers, and a distributed computing environment.

Infrastructure for Scheduling

The integration schedule was a risk. This resulted from several factors, not the least of which was underestimating complexity. Some individual systems lagged in meeting their own development and testing schedules, compressing the integration schedule further, and increasing the difficulty. The initial plan insufficiently accounted for the time needed to resolve the problems encountered in integration, and a high rate of changes resulted from the collaboration of operators and developers.

Initially there was no detailed integration schedule, rather individual and separate understandings of it. This led to difficulties in coordinating activities among the various teams, compounded by daily activities for SOS integration that were dynamic and changeable.

Project management tools were used to develop daily schedules, although their use was resource-intensive. Daily activities were posted, and a web-based approach was successfully used to disseminate schedules. It was essential to have contingencies available, and test directors at the integration site were prepared for schedules with many alternative plans. When one set of activities was prematurely aborted, others could proceed. Consequently, the trail boss was able to succeed in pushing the daily integration events forward.

Configuration Management Challenges

Configurations of individual systems were not always synchronized with those in the SOS baseline. This was exacerbated by the high rate of changes of the systems in the SOS, and because some managers other than those of the CTSF leadership controlled the baselines of some individual systems.

Teams at the CTSF sometimes found themselves in the situation of preparing for an event that had been rescheduled, or teams found themselves in the situation of testing their system in conjunction with another that had a different-than-expected configuration of hardware and software, a situation invalidating the test. Corrections to discrepancies would not always succeed because of hardware and software changes made elsewhere. Parts of the configurations were controlled, while others were not. Changes and reiterations of tests resulted.

There was insufficient opportunity and means to develop a suite of SOS test results and compile benchmark data sets for the operational threads through the SOS, such as from the battle drills and exercises at Fort Hood. This made some regression tests more informal than formal, sometimes with unexpected and undesirable consequences.

Conclusion of the System of Systems Integration Phase

While the LZ Phantom exercise in June 1996 signaled the start of SOS integration, analogous training exercises marked interim progress and completion. Platoon Lanes occurred from August through mid-September, Company and Team Lanes in October, and finally a Task Force/Brigade Team Field Training Exercise in mid-December 1996.

These built on the end-to-end scenarios postulated as to how a brigade task force would use the new digitization capabilities. The SOS providing these, though still perturbed by immaturity levels in some key systems needed for the experiment, was deemed sufficiently mature for the field experiment. The pieces of equipment were shipped from Fort Hood to the NTC in December 1996 to support the March 1997 engagement.

With leadership and commitment, the teams **behind the Wizard's curtain** coped. Their goal was to deliver a *"product mature enough for an experiment to provide data to the Army leadership for making investment decisions..."* (Boutelle, 1996), and in this they succeeded.

Chapter 5

Lessons Learned on Integrating a SOS

It is perhaps simplistic, but not overly so, to observe that the integration phase of both programs was to some extent an act of enlightenment. This is not meant to detract from the expertise and understanding of their program management and participants. Rather it is to observe that the complexity of integrating individually developed heterogeneous systems, managed autonomously, each complex in its own right, is staggering. The difficulty is well beyond that experienced in managing a single information system. It is the SOS entity that delivers the results required. It is not the individual systems that provide the necessary effects, although each contributes to them. The whole is greater than the sum of the parts and the value-added is achieved in the integration process. It is a strenuous undertaking.

The early stages of integration were confounding and overwhelming with the nature and number of problems, but both programs moved on. To the credit and success of both ventures, some adjustments were possible in methods and strategies, and the programs succeeded. If they had not achieved the integration, there would have been no DPS or TF XXI to speak of. There would have been only individual systems.

The lessons extrapolated from their experiences are presented in this chapter. They are also summarized in Appendix A. For future integrations, they provide a description of those participants who should comprise the team **behind the Wizard's curtain** and of the environment that should support them.

Summary of the Approaches to Integration

There were many similarities between the two programs in their approach to SOS integration. Neither organization previously had integrated the set of systems in the SOS. DMA accomplished *two* SOS integrations—for the DPS IOC and FOC milestones. Both prepared for the integrations through an elaborate and complementary cascade of testing, which began with the individual systems and interfaces, used independent review processes, and followed with another series of tests of interfaces by independent means. Both based their strategy on an analysis of risks, accomplished years[1] in advance of the operational need date for the SOS.

Characteristically they relied on testing events primarily, but not exclusively, targeted at individual systems and their interfaces. Both engaged external

[1] DMA assessed the integration risk at least 4 years before FOC, and the Army completed a risk assessment at least 2 years before TF XXI.

organizations for independent testing. Both used prototyping of individual systems for many individual system developments.

Both recognized the need to reduce risk with early integration efforts. The Army made extensive use of real and virtual simulations, deriving early knowledge about the behavior of core (to the experiment) systems and subsets of the SOS. The Army also integrated subsets of the SOS for use and experimentation in discrete exercises well before the SOS integration. DMA did this to a lesser extent but did integrate specific SOS subsets at the factories of individual systems to identify problems.

Both tested the integrated SOS in conjunction with operators. Both used a spectrum of end-to-end threads of operational activities to drive out problems in the SOS and signal completion of the integration phase. Their objective was to deliver an integrated product sufficiently correct for their needs,[2] which were different.

They were typical in their approach of "build-a-little, test-a-little" for the process of integration. As the end-to-end threads were exercised by operators, problems were identified and corrected. In the case of the DPS, the operational scenarios were demonstrations, including the use of digital source materials to produce specific MC&G products. In the case of the TF XXI, it

---

[2] The Army's objective was to achieve a product *"mature enough for an experiment to provide data to the Army leadership for making investment decisions..."*(Boutelle, briefing slide, 1996). For DMA, it was to acquire a production capability.

was the conduct of the LZ Phantom exercise and a series of battle drills and exercises at Fort Hood.

Both relied on one site as the place where the SOS would all come together for the first time. The DMA had planned for this with a new Production Center; the Army capitalized on the CTSF at Fort Hood.

Both were schedule-driven at the start of integration. The DPS was at the ending phase of a 10-year development, and DMA required the integrated capability to meet mission requirements. The Army needed the results of the TF XXI experiment for subsequently fielding capabilities as part of the Army XXI program. While both schedules were compressed, they were considered a reasonable risk.

There were some differences in their approaches. One is a similarity as well—the early integration of subsets of the SOS. The Army did this to a greater extent and planned the SOS integration incrementally, such as by components and systems, platforms, communications, and command and control. The DMA approach reflected the perception of the DPS as a digital pipeline, and progressed through the SOS integration of the systems serially through the various stages.

One primary difference in their integration strategies stems from the differences in their development methodologies. The Army defined an architectural framework and imposed it top-down on the individual systems. Careful attention was given to architectural compliance (and architectural evolution) as part of the testing strategy. As the SOS integration proceeded, this attention to compliance was sustained throughout the process. DMA accomplished the DPS architecture development bottom-up, within a far more generalized framework.

A second primary difference is that the Army used a spiral process of acquisition and development while the SOS integration continued. The DMA used a classic waterfall approach. For the TF XXI, the integration event was a discovery process of requirements and refinement of designs—iterating to modify the capabilities to accommodate the requirements as realized through the operators' experiences in conjunction with the developers who stood ready to respond. With respect to integration, this resulted in more impetus for changes in the individual systems, thereby affecting the dynamics of the integration process.

At the outset of the integration phase, the management of both ventures found themselves in similar circumstances. Both had underestimated the challenge of the SOS but moved forward to deal with it.

The Lessons Learned

Nine lessons learned are discussed here based on the two case studies and the influence of the integration environment on them. As articulated, the lessons learned appear fundamental. However, they serve as guideposts for dealing with the complexity of a SOS, providing practical strategies to supplement other good engineering practices. For future integrations, they provide a description of those personnel who should comprise the team **behind the Wizard's curtain**, the processes they should apply, and the common infrastructure they should use.

A Lesson on Preliminaries

Lesson 1

Certain activities should precede a SOS integration. These include:

defining the SOS architecture;

developing and testing the individual system constituents of the SOS;

developing and testing the interfaces between and among the individual systems of the SOS;

independently certifying compliance with the SOS architecture.

The promise of plug-and-play is a seductive one. Neither program presumed this advantageous circumstance, yet the difficulties of the integration and the complexity of the emerging entity far exceeded their expectations. This lesson serves to remind that verifying the individual constituent systems and interfaces and certifying them for architectural compliance are not activities which conclude the integration process but rather are necessary to *begin* it.

The lesson is applicable to a SOS which is a "new start" or one which incorporates many legacy systems. What both programs illustrated is that testing the individual systems and their interfaces is not equivalent to integrating and testing a SOS. Never mind that the constituents in each SOS were designed (e.g., DPS) or engineered (e.g., TF XXI) to function as parts of the cohesive whole from the outset. The whole is greater than the sum of the parts. In effect, a SOS integration process starts by using well-tested, certified individual systems and interfaces—and an architectural framework.[3]

For the two case studies, all the earlier development and testing—the prototyping, testing of individual systems, the independent verification of interfaces, and the integration of subsets—were necessary to deliver functioning *individual* capabilities to the integration site. But the functioning of individual systems is not equivalent to the functioning of a SOS. Previous activities like independent certification helped to resolve a myriad of problems of interoperability and non-compliance with the architecture. However, in the aggregate these activities were not sufficient of themselves to achieve integration of a SOS. The extraordinary efforts[4] of an integration process were required to accomplish that task. The integration process was necessary to complete not only the "plugs" but establish the "play" for each SOS. Even when the interoperability between and among the constituent systems is readily achieved, the integration process is needed to ensure that the SOS provides the results required.

---

[3] The architecture may convey common applications used by constituent systems, such as from the DII/COE.

[4] Chap. 4 describes these in detail.

Both programs did encounter an increase in difficulties because some individual systems were not mature before, during, or even after the SOS integration. Many individual systems were themselves cutting-edge developments and therefore at different levels of maturity. Both programs progressed to produce an integrated product, but its quality was lower in early use than later use. Typical impacts included discoveries of new discrepancies, interruptions or delays in services, and the down-time of components.

A Lesson on Continuous Integration

Lesson 2

Use early, incremental, and iterative integration to achieve a SOS.

In the acquisition cycle, the more traditional timing for the integration phase follows the development and testing phases of all the individual systems and interfaces of a SOS. There are benefits to beginning the integration of a SOS earlier than this classic methodology. There is risk in dealing with complexities and unanticipated behavior late in the cycle. The difficulty can be so great as to result in failure to deliver—or to deliver the integrated product extremely late.

An additive suite of events, including early use of simulations and exercises, and early integration of systems and subsets of the SOS, enhances the preparation for SOS. This is less risky than approaching the integration as a one-time "big-bang" event when all the constituent systems are available. SOS integration is a resource-intensive process of build-and-test and should be planned accordingly, preferably in iterative increments. It is not a simple process of snap-in, snap-out. It is a difficult undertaking in the best of circumstances.

The experiences of both programs, as well as good engineering practices, argue to tailor the strategy not only to early, but also to an incremental, iterative approach.[5] This means beginning the integration of a SOS with a subset of systems, then reintegrating these with additional subsets of systems in a series of iterations, until a last phase of integration of all the systems is completed. Early and iterative integrations enable a more systematic accommodation of changes, and allow sufficient time in the program venture to adapt when the unanticipated occurs.

A first-time integration occurs when *all* the systems of the SOS have never been combined in their then-current form. Both programs produced the first example of a SOS because they had not integrated all the systems previously.

The Army exercised a strategy that relied on integrating *subsets* of the TF XXI SOS. Yet the program still experienced considerable difficulties when all the systems were integrated the first time. The immaturity of some components did not always allow systematic increments. Difficulties also resulted with their

[5] For example, see Rechtin and Maier (1997) on stable intermediate forms and Walker Royce (1998) on iterative life-cycle processes.

incremental approach because changes continued to components, systems, and subsets after their earlier integration. The architecture incorporated constituent systems that had been integrated previously to support earlier experiments. However, these systems evolved in the interim. For the TF XXI, they were then combined with new and different digitization initiatives, resulting in a new integration challenge.

A second integration event of the DPS occurred for FOC—started less than a year after the first. This reintegration manifested all the difficulties and complexities equivalent to that of the first-time integration because so many changes had occurred. The SOS was comprised of the same set of individual systems but their functionality and interfaces had all changed in non-trivial ways.

Both external and internal forces will propagate changes during the acquisition of a single SOS. The cautionary note warranted for a SOS integration is connected with the effects of changes to systems or components already integrated—as illustrated in the two case studies. When change occurs, consequences occur for reintegration. A SOS, which characteristically is comprised of constituent systems that themselves are large-scale, can be so complex as to behave as a non-linear system. This means that small changes can produce disproportionate results; the whole is more than the sum of the parts; behavior does not always repeat; and causes and effects may not be demonstrable (Czerwinski, 1998).

With the strategy of iterative and incremental integration, the question arises as to how to begin. Generally good engineering practices argue that the approach should be a risk-based one (i.e., taking the systems or components that are most difficult and risky to integrate and focusing early-on to resolve issues). This is certainly consistent with DoD acquisition practices (Schaeffer, 1998). If the systems fail to mature or the issues are not resolved within the allotted time, there is more time to make adjustments.

Early integration also should focus on core capabilities needed for the mission or experiment. While it is critical to "do the hard parts first,"[6] it is equally important to concentrate on the parts that matter. Progress allows secondary objectives to be introduced in later iterations or subsequent phases of the program venture. Because success in integrating a SOS is a strenuous undertaking, narrowing the focus to core objectives is a practical strategy. Both case studies provided examples of requirements (e.g., centralized program management in DPS) or initiatives (e.g., in TF XXI) that migrated beyond those that were essential, complicating the integration. Many ventures so encumbered will not succeed.

The argument for early integration also aligns with recommendations from experiences incorporating substantial numbers of COTS products into information systems and enterprises of heterogeneous systems (Fox, et al., 1998; Brownsword, et al., 1998):

---

[6] Rechtin and Maier (p.42) use this as a heuristic.

> *...in today's reality, software COTS products seldom plug into anything easily. Most products require some amount of adaptation to work harmoniously with other commercial or custom components in the system.... Adaptation must take into account the interactions among custom components, COTS products, any non-developmental item components, any legacy code, and the architecture, including infrastructure and middleware elements* (Brownsword, et al.).

To deal with the rapid turnover of COTS products, recommendations include an immediate up-front integration and test activity with other systems and components, as well as construction of an environment to do so (Fox, et al.).

An early start and the use of the incremental iterative approach, while not eliminating the considerable complexity of the undertaking, serves to reduce the risk and apportion the difficulty into more manageable segments. It also provides insight into the holistic behavior of the SOS earlier and incrementally.

A Lesson on Testing Strategy

Both programs recognized the need to use operations-like activities typical of the mission to build and test the integrated product. They also recognized the importance of using all the actual systems in the SOS at the integration site (in contrast to heavy reliance on synthetic stimuli), for the final integration of the SOS.

Lesson 3

The testing strategy for the integration of a SOS requires:

an agreed-to plan and process for testing, based on a risk assessment;

a suite of activities representative of the operational requirements of the mission the SOS supports;

the exercising of a full spectrum of the SOS activities (end-to-end) by operators, using the actual constituent systems of the SOS—or at least a core SOS.

This lesson synopsizes what *is* required. Anything less is a compromise and elevates risk. One of the challenges attendant to integrating a SOS is understanding its behavior sufficiently to make informed decisions about a compromise. Earlier integration (and testing) of subsets of the SOS serves to reduce the risk, as well as to establish patterns of operational usage and SOS behavior. These provide information needed to devise and refine the testing strategy.

Each SOS integrated in the case studies was highly operator-interactive[7]—in contrast to one deployed on a space-based platform. When this is the case, test plans need scenarios built by operators and exercised by operators. Both programs developed test plans using end-to-end scenarios. With limitations on time (there are always limitations), the challenge lies in assessing the sufficiency of the end-to-end threads and of the risk incurred by a compromise in testing with less. Both organizations were influenced by the demands of schedule but judged the risk reasonable.

The DPS demonstrations included end-to-end mission threads, but not for *every* mission nor *all* products. The TF XXI used a set of battle drills and exercises. The sufficiency of these for completeness of integration testing is vulnerable to the ingenuity of the operators and how accurately the set reflects actual operational scenarios. Also the nature of operational use of a SOS evolves. In both case studies, problems were found after integration when the unexercised threads were exercised.

In addition to early integration of subsets, a useful strategy to assess completeness of the testing is to analyze defect trends. There are methods to predict the number of defects in an information system.[8] Assessing the number and nature of discrepancies detected and corrected provides information on errors remaining. A decline in problems uncovered by operators over time should signal growing stability of the SOS.

During the SOS integrations of both programs, there were also compromises when certain systems or components were unavailable; simulated capabilities were used.

With SOS complexity, a better guarantee of sufficiency of testing is to use the actual systems. Models and

---

[7] Operational missions will rely on a human-interactive SOS because warfare is based on human behavior.

simulators are, after all, only an approximation of the real thing. However, they are important *complements* to testing methodology.

A core SOS brings significant benefits for integration and operations. DMA successfully constructed such a miniature DPS—called the System Test Mode (STM)—dedicated specifically to verification.[9] This was used for the systematic assessment of changes in the integrated SOS baselines while evaluating the results with an increasing number of benchmarks. This enabled resolving adverse side effects before operational use and verifying that changes did indeed deliver the results intended, both objectives important for maintenance and evolution.

A stand-alone core SOS can be exercised concurrently to the SOS being used for operations. As such it should be physically isolated from that being used in actual operations, including independent data bases. Changes to components and systems of the SOS then can be exercised with full benefit of regression testing and the use of end-to-end test suites. For DPS, when it was difficult to include certain unique components in the STM, simulators such as for the imagery network were used. However, such substitutions frequently introduced anomalies of their own into the verification process.

A core consisting of a minimum set of capabilities may be inadequate to verify performance of the integrated product under peak stress conditions—when the

[8] At least three methods were used for the DPS, including one calibrated for large-scale DOD models.

[9] See chap. 4, section entitled, "A DPS Core."

maximum is needed. In this circumstance, other means, such as modeling and simulation, may be the only or best supplement.

A Lesson on Planning Resources

Lesson 4

To integrate all the systems of a SOS, plan for substantial difficulties and significant time and resources.

This lesson rejects outright the assumption of plug-and-play, even while presupposing that the appropriate testing of individual systems and their interfaces has occurred. It provides a counter-assumption for planning purposes based on current experience—that the integration of a SOS is a strenuous undertaking. As such it does not imply faulty testing of the individual systems. Rather it underscores the efforts that must be expended to achieve the emergent and required behavior of the SOS.

The TF XXI integration phase at the CTSF at Fort Hood required 6 months. Generally activities occurred around the clock 7 days of the week. Approximately 500 personnel supported just the integration activities.[10] For the DPS, the first formal integration event required 7 months and the second required 10 months. Activities also occurred around the clock every day. About a thousand personnel supported the DPS integration events.[11] For both case studies, factors contributing to the resource requirements were that all systems were not previously integrated and that many constituent systems were cutting-edge.[12]

A SOS is a complex venture, with individual systems typically in large scale. For military missions, advanced and cutting-edge capabilities often are inserted to provide an operational advantage. For many operations, forces and systems are combined in unanticipated ways. Planning for substantial resources for a SOS integration **behind the Wizard's curtain** is warranted. The resources, if not properly programmed a priori, can be problematic to acquire because large expenditures are needed.

Planning requires good estimation methods, as well as a good experience basis for estimation. While this lesson points to the need to be prepared for the ravages of integration of all the SOS systems, developing credible estimates of how much time and resources are necessary is challenging. Both programs used methods to determine the time and resources needed that proved inadequate in light of the actual effort required.

---

[10] This estimate included personnel at the CTSF who supported the processes and infrastructure for integration, and factory personnel who modified and corrected constituent systems and interfaces for the integrated product. This number does not include operators.

[11] The number was less for IOC; it was more for FOC because the latter milestone affected three Production Centers. The DPS estimate includes the DMA personnel supporting integration and transition to production and contractors supporting integration and correcting constituent systems and interfaces at factories. The number does not include operators.

[12] After the DPS FOC and the TF XXI event at the NTC, reintegration of the SOS required fewer resources as constituent systems stabilized and changes slowed.

Methods for estimating time and effort for complex information system projects continue to evolve to provide the means to factor in complexity, modern methodologies, and larger-scale systems.[13] Today's program manager has many such models available for estimation. Nevertheless, while systematic measurement should be applied, current estimation methods generally are based on software projects or information systems. Because a SOS involves constituents of large systems independently managed, developed, and operated and its behavior is emergent, and because the venture is resource intensive, more accurate means are warranted. The need for accurate estimation is compelling for an unanticipated real-world military operation, where it is essential for successful mission planning.

To improve this situation, estimation models need tailoring and calibrating using actual expenditures from SOS programs. This will be discussed in chapter 7 as future work.

A Lesson on the Importance of an Integration Facility

The use of a single site and its supporting environment for SOS integration was essential in achieving success. In an era of digital communications, the transport of faces and data can be accomplished in seconds. But a virtual team cannot sustain the empathy nor the

[13] For example, see discussion on Cocomo II (Boehm, et al., 1996).

continuous interactions necessary for the birth of a SOS. In addition, the SOS did not exist until it was integrated.

The collocation at the single facility of program managers, engineers, and users enhanced the understanding between developer and operator to the benefit of all. The single site environment facilitated team work focused on the success of the SOS (as compared to individual systems). It enabled more judicious resolution of problems in the shortest possible time. The integration site was used by both programs as a means to maintain schedule after underestimating the complexity of the SOS integration. It enabled the teams to form the parts into a whole. Finally, the SOS capabilities "born there" were necessary to support the training of the operational community in the use of the SOS, an aspect discussed in chapter 6.

So critical for the success of the TF XXI was the CTSF environment that the Army subsequently expanded its use for other integrations, including the TF XXI Division Exercise.

Lesson 5

The use of a single facility—with an environment of people, processes, and infrastructure—substantially facilitates the integration of a SOS from individual systems.

The value of a single facility with its supporting environment was clearly demonstrated in these two case studies. Their experiences showed the following effects:

- More effective leadership of the SOS integration resulted from the collocation of management and engineering resources from the constituents.
- A SOS team emerged where previously there had been individual teams.
- The holistic behavior of the SOS was constructed dynamically at the site through the collaboration of those individuals responsible for the SOS, those with the diversified, specialized understanding of the individual systems, and the users.
- Standardized processes and infrastructure in the environment enabled effective, efficient, and unambiguous exchange of information.
- Complex information to identify and resolve issues was communicated and comprehended dynamically and continuously with necessary personnel participating.
- The integration was achieved within the allotted schedule.

How, or if, distributed integrations of a SOS might be accomplished successfully is discussed in the context of future work in chapter 7.

A Lesson on "Who's in Charge"

Steps were taken at the outset of both program ventures to ensure that the overarching authorities and responsibilities for delivery of a SOS were clear. Ambiguity or lack of clear decision-making authority was not a problem in either case study. Rather, the clarity of lines of responsibility contributed to the success. The acquisition responsibilities were simplified in that primary accountability rested with a single program executive officer (PEO) within a single organization. DMA's PEO controlled the resources. In the Army's case, multiple PEOs controlled multiple pools of resources, but the accountability was allocated to one, the PEOC3S, for the TF XXI experiment. The operational responsibilities were defined clearly for both programs. Steering groups and working groups for the participating organizations provided direction, coordination, and recourse for problems and issues as necessary.

In both cases there were also clear lines of authority delineated specifically *for the integration and at the integration facility.* A cadre was responsible for the SOS and was distinct from those teams managing the individual systems of the SOS. The experiences lead to the following lesson:

Lesson 6

The process for SOS integration should overtly address the leadership of the integration as follows:

an on-site acquisition leader empowered for the integration of the SOS and an on-site leader empowered for the operational community;

supported by a SOS cadre—with sufficient resources and authority;

supported by participants who manage, develop, and operate the constituent systems of the SOS.

The question, "who's in charge of the SOS integration," was addressed openly by the management of both programs. The DMA appointed a senior leader-engineer at the integration site who reported to the DPS PEO, and the Army designated a trail boss acting on behalf of the PEOC3S at the CTSF.

Also demonstrated was the importance of an on-site leader empowered for the operational community. The SOS and the individual systems are in a dynamic state during integration, requiring decisions that must be closely coordinated between the development and operational communities. An authority to speak for operations at the integration site preserves a uniform operational view and agreement on operational priorities. This is particularly important because requirements will be changed, refined, introduced, and deleted during SOS integration. Presence at the site provides first-hand understanding of the operational experience with the integrated product.

A SOS cadre was established by both programs, and among their duties was direct support to the on-site leaders of the SOS integration. The case studies illustrate that they required more resources, painfully evident at the time of SOS integration. For both DPS and TF XXI management, realization of the level of resources required and the extent of expertise needed emerged fully at the time of integration. The problems encountered in the early stages of SOS integration resulted in immediate reactions by the leadership of both programs to increase these assets.

The SOS cadre has duties that are distinct from those of the program managers of the individual systems. They are required for more than the SOS integration event, in fact for all phases of a SOS undertaking—such as in evolving the specific SOS architecture. As such, there are advantages to retaining these same individuals throughout all the various stages of the life cycle of a SOS.

The experiences may indicate that the peak level for their resources occurs during the SOS integration phase, and both sets of events reveal the difficulties of introducing new resources late in the process. But more data are necessary.

The cadre must have power and influence in the decision processes, both acquisition and engineering. Current acquisition processes recognize the roles and responsibilities for program management of individual systems. In both the DPS and TF XXI ventures, the program managers of individual systems were perceived as carrying greater responsibilities than the SOS cadre. They managed large acquisitions, many people, and controlled significant dollars.

Substantial adjudication from both the programmatic and engineering perspectives was required on behalf of the SOS. Decisions had to be made favoring the whole SOS entity at the expense (literally) of the individual systems. This of itself illustrates the need for a cadre with leadership responsibility and authority for the SOS.

When the SOS integration required additional funding, adjudication across these various interests was required. Consensus or negotiation is not always the most expedient process. Greater autonomy and heterogeneity in management teams are more representative of future SOS ventures. This implies the need for adequate staffing and more influence in the SOS cadre, plus control of funding by the SOS integration leadership sufficient to deal with unanticipated problems in the integration.

Program managers of the individual systems, and their development teams, as well as users who operate these systems for integration testing, are also essential members of the "one team" engaged in activities **behind the Wizard's curtain**. Collectively they are responsible for the "parts" that comprise the SOS but also support missions of their own. They provide the needed engineering and operational insights into these constituents, and accomplish what is necessary to develop or adapt, integrate, and sustain those systems in a SOS.

A Lesson on Common Processes and Infrastructure

While there were some differences in common infrastructure at their respective integration sites, there was a correlation on many essential processes for the two case studies—engineering boards, configuration management, and external communications.

This lesson communicates the need for certain common processes and common infrastructure in the integration environment that are the same for all teams. It is likely that the organizations that develop and operate the individual systems of the SOS use different processes, tools, and infrastructure to manage their own developmental and operational capabilities. It is expected that they will continue to do so. But for the SOS integration environment there is commonality that is minimal but sufficient—and available at the integration facility.

Lesson 7

Certain common processes and common infrastructure in the integration environment are essential to manage a SOS integration successfully. These include the following:

an Engineering Board with responsibility and authority for identification and resolution of SOS issues and discrepancies, including the assignment of responsibility for correction;

establishment of processes (and the automated means) for identification of SOS issues and discrepancies, their disposition, tracking, and resolution, under the management of the Engineering Board;

automated support for the tracking and tracing of SOS operational requirements;

configuration management and control of the hardware and software baselines of the systems of the SOS by the integration leadership, supported with: automated means for identifying and controlling the baselines and subsequent changes; a formal build, verification, and re-integration process for changes;

a robust communications infrastructure linking the teams internal to the integration environment and their external counterparts;

an office automation environment to support the integration's administrative processes as well as to support interpersonal processing and communications for the participants.

There is a requirement for the integration leadership to designate the actual tools[14] and infrastructure that constitute the common integration environment for the SOS being integrated. Early specification is important,

[14] For example, designating a specific tool for hardware and software baseline configuration management enables teams that manage the individual systems (and use other configuration management systems) to plan accordingly.

as the DPS case study demonstrated in the example of the common discrepancy (defects) tracking system.

A subtle but important nuance on configuration management processes is articulated in this lesson. The configuration management of the individual systems in the SOS should be controlled by the SOS integration leadership, *not* by the management of the individual systems. This arises from both sets of experiences. DMA was able to substantiate the benefits of this approach on its second DPS integration.

Both programs exercised configuration management of the baselines of the individual systems during SOS integration. However, for some specified systems they allowed the primary change control to be managed by the factories of those systems. These exceptions produced problems. As a check on process after SOS integration, DMA re-exercised the end-to-end threads with the integrated product and identified anomalies in the individual systems' baselines. Consequently, during the second SOS integration, the primary control of all the individual systems was allocated to the SOS integration leadership and no exceptions were allowed. This resulted in a more robust process for change control.

Because an individual constituent system is likely to be independently operated for multiple purposes (other than that of the SOS being integrated), it may require management of multiple baselines. The change control of the baseline in the integrated product should be managed by the integration leadership. A plan and process for subsequent synchronization is then required.

Consistent with the dynamics of change, configuration management tools used should be sufficiently robust to allow rolling backward and forward through many versions.

Robust communications infrastructure is essential to link the developers at the integration facility with their factory counterparts. Linking the integration site to locations where the SOS is operated also supports its subsequent deployment, sustainment, and evolution.

A Lesson on Efficiencies and Effectiveness

There is a need to optimize use of the large resources attendant to a SOS integration, and to preserve schedule. Both programs relied on daily status, on alternative and contingent plans, and the dissemination of program information. Because the constituents of a SOS are independently operated, developed, and managed with their own individual processes, it is beneficial to establish these methods and automated infrastructure as part of the common integration environment as early as feasible.

Lesson 8

Certain common processes and infrastructure in the SOS integration environment promote effectiveness and efficiencies. These include:

daily planning and scheduling of resources (people, equipment, facilities) for integration events—with contingency plans and schedules readily available;

timely dissemination of information pertinent to each integration event, such as test status, equipment availability, and results;

daily status meetings, with results immediately available.

Daily planning and scheduling tools aid the integration management in optimizing participants' time and energies while moving the overall program schedules forward. The DPS integration leadership used such capability extensively to manage the 700 daily events that were ongoing for SOS integration, affecting more than 1,000 people.

Such tools facilitate progress through the integration schedule. While automated means can project the daily events, the dynamics of SOS integration often preclude them from taking place as planned. Automated tools should provide the means to develop alternative or contingent plans and link related activities. For both programs, being prepared with such alternatives for the multiple daily events ensured that progress continued.

Both programs experienced considerable contention for resources—people, equipment, and facilities. Using automated planning for the daily events supported deconfliction and optimization of resources.

The immediate availability of status was important to gain a common understanding by all participants at the same time, as opposed to individual, disparate, and unsynchronized understandings—which carry the consequences of unnecessary, uncoordinated, and futile activities.

A Lesson on Evolutionary Acquisition

Lesson 9

Prototyping a SOS can provide early insight into operational requirements and into the SOS systems architecture.

Generally this lesson encapsulates the merits of prototyping, which can be applied to any new software project or any new system. However, articulating the lesson here emphasizes the distinction that it is the SOS that delivers the necessary results—not the individual systems. The performance of the individual systems must be understood in the context of their contribution to the SOS behavior. How the whole entity behaves is the crux of the matter. In turn, all the systems in the whole must be integrated to realize the full extent of the holistic behavior.

It is the integrated SOS that provides the insight as to whether the delivered capability renders the results expected—and required. However, the results also may propagate a different understanding of what is needed—and imply changes in the operational architecture. When using experimental capabilities to support experimental concepts, the key is to use the results of integration as a contributor to the requirements and architectural processes for the integrated product intended for field production.

For each case study, the SOS provided revolutionary capabilities and supported revolutionary operational concepts. The entity emerging from the integration of the individual systems was difficult to predict at the outset, a theme frequently played within this work.

The Army integrated a specific SOS for the first time to put digitization capabilities in the hands of the operators for the 2 weeks of the TF XXI at the NTC. As a result, the Army gained considerable understanding of the consequences of that capability, and from many aspects—operational, doctrinal, and structural. The information derived was used for many purposes—including follow-on acquisitions for the Army XXI program.

The Army viewed the acquisition process that occurred at its CTSF as one of the fundamental successes of the TF XXI experiment. The process that ensued was broader than just prototyping—it was one of evolutionary acquisition (Boutelle & Grasso, 1998). The collocation of developers with operators training on revolutionary capabilities produced a spiral of feedback on requirements and refinement of capabilities, trimming years from the usual acquisition cycles. This advantage cannot be overlooked as an important benefit of the SOS integration environment—and a future opportunity.

DMA elected to forego prototyping the DPS[15] as a strategy in an era where the more classic waterfall methodology prevailed, and one can only speculate as to the impact of a decision that appeared sound at the time. After FOC, the integrated product provided new understanding, which resulted in changes to the initial operational concepts, such as those for production management. Without insights into the behavior of the SOS earlier in the process, the surprises can only be greater, considering the complexity of the undertaking.

While the integration environment can support management's ability to move forward after underestimating SOS complexity, it cannot overcome the shortfalls of bad requirements or a poor design, or be a substitute for good architecture. However, integrating a SOS as a prototype for experimentation can provide significant insights into the consequences of operational requirements and design decisions earlier in the acquisition process, and it is an alternative to a methodology that is more serial, rendering that understanding near the end of the acquisition cycle. It is also far better than relying primarily on the knowledge of the capabilities and performance of the individual systems.

The caution that must be sounded is that, in accordance with the Heraclitan principle of change, and consistent with the experiences of both case studies, the integration of the SOS capability ultimately fielded will still be a strenuous undertaking.

---

[15] The concept of a Mark 87 engineering prototype of Mark 90 was considered briefly, but subsequently terminated.

Chapter 6

Lessons Learned on Training with a SOS

This chapter discusses the training experiences for both programs. Training and the SOS integration occurred concurrently. Because of the desirability to train operators on capabilities identical to that used for operations, the integration environment was essential for training. The integrated product was needed to accomplish the mission, and the SOS for each program was incrementally evolving at the integration facility through the "build-a-little, test-a-little" process of integration.

Three lessons learned about training with a SOS are discussed in this chapter. What emerged from both sets of experiences was the importance of allowing more and iterative training on the integrated SOS capability—and the need to teach the whole in addition to the parts.

Training lessons are summarized in Appendix A.

The Digital Production System Training Program

For the DPS, the training program focused on the Agency production workforce. In addition to managers, this workforce included primarily cartographers and those in related fields. The total training population was about 2,700 people.

The DMA management provided training on the individual systems of the DPS, coupling these with prerequisite and related courses (i.e., computer skills and knowledge engineering). Generally the training program assumed that students knew their profession and focused on the new digital capabilities with new operational scenarios. Multiple training events prepared the workforce for the daily business of using the DPS to produce MC&G products. More than 8,000 training events were conducted.

The most challenging training events for operators were the Exercises and Rehearsals (E&R). These consisted of rehearsals coupled with exercises to produce a set of DMA products end-to-end using the integrated SOS. These events were conducted first at the Production Center that served as the integration facility, then expanded to the other two Production Centers. The set of products included those produced in the integration demonstrations and others as designated by production priorities, which varied over time. The schedule serialized events to engage operators in increasingly comprehensive production activities with the new capability, while training operators in time for their production assignments using DPS. The events for E&R required many months—consistent with the long production time required for Agency products, even with the new capabilities. The timelines were also deliberately lengthened to allow for learning. The training window extended well past FOC to accommodate the large training population. Figure 6-1 illustrates the training schedule embedded within that of the DPS program.

Early in the series of training events, operators exercised the functionality of an individual system in the standalone mode, while training for their part in the new DPS operational concepts. An operator on the system for data extraction learned to apply the functions of the newly developed workstations—how to perform feature extraction for multiple products, and how to populate the MC&G data base. Both multiproduct extraction and attributing the data base were innovative concepts introduced by the DPS.

The operators' initial access to the integrated suite of DPS SOS hardware and software occurred during the demonstrations before IOC, but this was only by a small cadre. During the E&R and FOC demonstrations, the number of users increased substantially as 20 percent of the workforce was trained. Their experiences were used to generate production procedures for subsequent use. After FOC, some operators helped train the remaining 80 percent of the workforce for DPS production while the majority moved to production assignments.

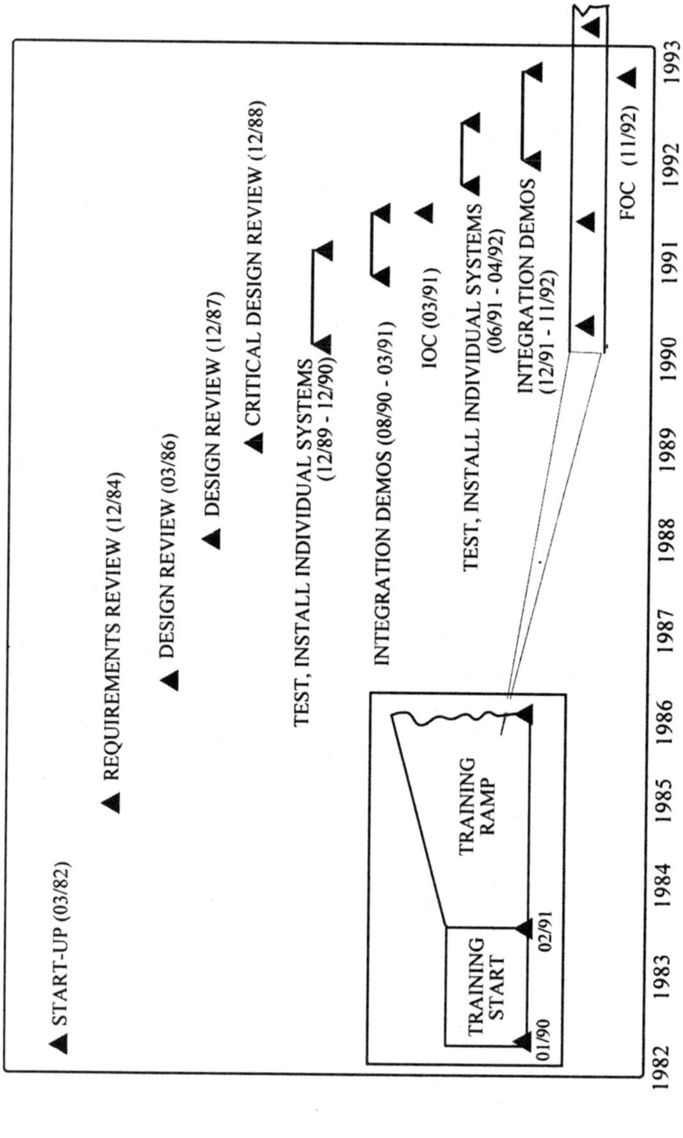

Figure 6-1. DPS Training Schedule

The strategy emphasized training an operator for the individual system he or she would operate in production; an operator was not trained on multiple systems of the DPS. Only general information was provided about the SOS capabilities and other (than the operator's) constituent systems. At the time, this was not perceived as a limitation. One training program intended for production managers attempted to provide a detailed exposition of the SOS and all individual systems and interfaces; however, it quickly became obsolete. Only a small population of the workforce attended.

In E&R, operators exercised production *tasks* analogous to their future work assignments When using their own workstations, such as to accomplish data extraction, they also engaged the support of other systems, such as those providing imagery and ancillary source information, and those managing the MC&G data bases. Yet the DPS operational concepts and the operators' training materials included limited information about these relationships. Materials did not provide the rapidly evolving and detailed "as built" interfaces; consequently, operators obtained only a top-level understanding as to how the processes and methods of other systems in the SOS impacted them in doing their own job. In reflecting on the training experience later, one senior DMA leader noted the rarity of people who understood the end-to-end process of making a map with the DPS, and the value of those who did.

During E&R and early production start-up with DPS, it was difficult for individual operators to understand what, why, and how functionality was executed beyond the boundaries of the operator's own individual system. The results of an individual's actions, which created ripple effects in the mapping pipeline, were not addressed in the training program. These effects were dynamic, changeable, and often non-repeatable. At times these experiences were inexplicable to an operator. The understanding of how an individual's actions related to the whole SOS was elusive.

The DPS underwent numerous corrections during E&R, primarily resulting from problems with interfaces during integration. Despite this, training continued as long as workarounds[1] could be developed. Training concurrently while the SOS was stabilizing was burdensome to operators, who were frustrated by delays, errors, changes, and rework.

Despite the objective to make E&R just like production, the full complement of equipment needed was not always located at the integration facility nor available for training. Some pieces were unique, such as scanners or plotters, and were not allocated to the integrated training environment. Training carried a lower priority than integration needs and than production needs after FOC.

These omissions resulted in training cartographers on abnormal configurations, which then introduced other anomalies. Because there were different production

[1] Workarounds were temporary procedures that deviated from normal production processes but allowed recovery from a functional defect. Workarounds were not intended to be permanent, but they gave developers time to correct discrepancies.

assignments and unique subsystems at each of the three Production Centers, configurations supporting the training baseline varied. The result was that operators were not always able to train on an environment identical to the one they used for production, and they experienced atypical problems.

More equipment was moved into production and unavailable for training events, and greater use was made of simulators to provide interface stimuli. This, too, created abnormalities. As more elements migrated to production, the E&R concluded earlier than planned. The Agency decided to accelerate production with the DPS to maximize its benefits. Nonetheless, the increasingly synthetic nature of the training environment decreased its effectiveness in preparing operators for production.

Digital Production System Training Program Well Received

Overall the training program was well received by the operator community; it exceeded a 90 percent acceptability level by operators and managers. When softcopy training materials were available at individual processors or operator workstations, they were found to be more effective. DPS operators preferred embedded training with self-paced instruction. In the opinion of trainers and trainees, this approach produced superior results. This level of embedded training was developed in only one individual system of the SOS, and it became a model for future training programs.

The Task Force XXI Training Program

In fact, TF XXI was an experiment coupled with a training event intended to provide an experience as close as possible to an actual conflict. The operator cadre was the Experimental Force (EXFOR) comprised of approximately 5,000 soldiers of a Brigade Task Force. The training addressed not only the digitization capabilities but also new tactics, techniques, procedures, and organizational initiatives.

Training for earlier digitization efforts, including AWEs, exercises, and simulations, had provided several lessons (i.e., that units should be proficient on combat fundamentals, be proficient with the digital systems, and train using these systems) (TRADOC Integrated Report Annotated Briefing Slides, 1997). In the year preceding the TF XXI, there were platoon- and company-level field training events; however, simulation networks were used as the principal means for battalion-level training.

The EXFOR's training on new equipment ramped in March 1996 following that of a small cadre of soldier-trainers who began 2 months earlier. A rigorous schedule was enforced because the training requirements were so vast. The equivalent of a "digital university" was established to enable soldiers to learn to operate one or several pieces of equipment. More than 90 classrooms and 28 motor pool areas were used for hands-on training (Goedkoop, 1997).

When most of the individual systems arrived at Fort Hood in June 1996, integration difficulties occurred, as discussed in chapter 4. However, training continued around the clock. Developers worked at correcting or adapting individual system capabilities based on operators' feedback in training. A continuous process of train, refine or correct, and train, occurred with operators influencing the integrated TF XXI SOS capabilities to reflect the realities of operational needs.

As the stability and completeness of the SOS integration advanced, a series of training exercises called battle drills were conducted with the EXFOR. These training activities used configurations supporting operational enclaves, like the Tactical Operational Command Centers. They were used in conjunction with other exercises at Fort Hood to enable the soldiers to operate with discrete subsets of the SOS as well as the integrated product.

According to Boutelle and Grasso (1998):

> *The development and execution of battle drills was key to the success achieved in training the warfighter. These battle drills simulated specific threads of operation and allowed the warfighter to better understand the tools and capabilities available in the context of his or her mission.*

Exercises were conducted at Fort Hood to provide phased training in a realistic field setting. Soldiers advanced from using basic connectivity to using increasingly comprehensive capabilities and applying new tactics and organizational initiatives. The training events are highlighted in figure 6-2, relative to the schedule for the TF XXI architecture and integration. The platoon-level exercise in August 1996 provided unit experience in a tactical environment. Company Lanes during October trained soldiers using the digitized combat support and combat service support capabilities. A brigade-level exercise, concluded in December 1996, gave all soldiers of the Brigade Task Force their first opportunity to maneuver together in a realistic environment using the SOS with digitization initiatives and legacy systems fully integrated.

The field exercises were supplemented with simulation-based training events. A classroom facility, equipped with surrogate appliqué workstations to simulate situational awareness, was used to enable EXFOR commanders and staffs to engage in war games with specific objectives. Simulated GPS data and certain intelligence assets, along with tactical vehicles, also were networked. This allowed exercising tactical decision-making processes and supplemented field exercises. When the equipment was shipped to NTC in late December, the EXFOR trained with battalion- and brigade-level simulations in January 1997.

Training experiences occurred on capabilities in flux, the result of constant corrections to individual systems, refinements induced by operator feedback, and progressing stability of the integrated SOS. New trainees used the capabilities differently than their predecessors as more changes and corrections were made. Engaging the operators in the exercises fueled these dynamics further by identifying more problems

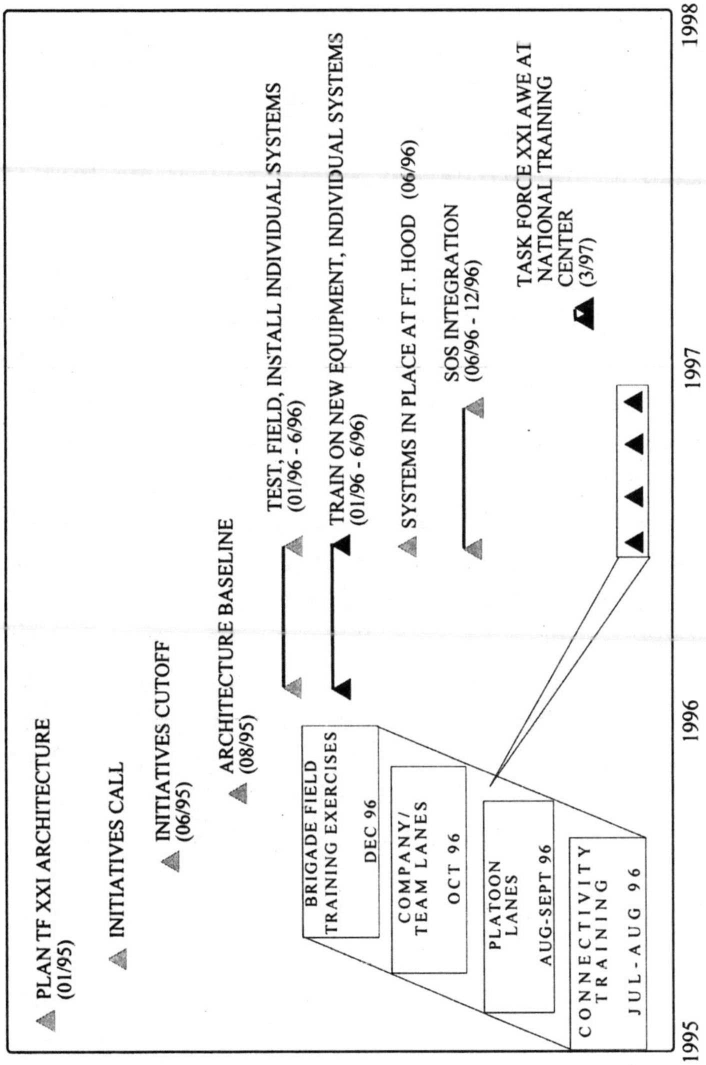

Figure 6-2. Task Force XXI Training Schedule

for correction, more user modifications, and more changes to the integrated product.

As a result, operators trained on an evolving, constantly changing SOS, though one better aligned with operational requirements. Interruptions were burdensome to the training experience; and those who trained early found themselves using altered capabilities during the actual NTC event.

A high rate of turnover in soldier participants occurred early in the EXFOR training cycle. Despite early efforts to sequester specific assignees to the EXFOR and retain them until the TF XXI experiment concluded, the implementation of this strategy lagged. Many soldiers who received early classroom training were reassigned. Between January 1996 and the AWE at NTC, personnel turnover, including duty changes, exceeded 40 percent. When the TF XXI occurred at Fort Irwin, many EXFOR participants did not have full training on current versions of SOS capabilities because of turnover and changes. For example, despite at least three major upgrades to the appliqué during the training period, only about half the soldiers previously trained received updates (TRADOC Integrated Report Annotated Briefing Slides, 1997).

Conclusions about Training with a System of Systems

The training strategy for both programs appropriately emphasized operators' access to new capabilities using operations-like scenarios. Both programs used a complement of training events. The TF XXI and DPS programs recognized the need to train for the operational mission in a fully integrated SOS environment and provided this access. This training was *in addition to* that provided on individual systems and components.

The results illustrate that the operational community was able to perform the mission required. Nonetheless, more and different training opportunities were necessary to *master* the SOS capabilities. The Army's special training assessment found the TF XXI training program effectiveness was affected by various factors, including immature equipment, a high rate of changes, and turnover (TRADOC Integrated Report Annotated Briefing Slides, 1997).

In training on the integrated SOS, both cartographers and soldiers experienced interactions that were different from training on an individual system in a standalone mode. This situation arose from three principal causes—interactions arising from the holistic behavior of the SOS, changes deliberately introduced because of user feedback (in the TF XXI case), and problems remaining in the integrated SOS. All prompted responses from operators for actions and decisions that had not arisen from training in the standalone mode or in a less than fully integrated mode.

From the perspective of "train-as-you-would-fight," this is problematic. However, it should be expected with a SOS.

The concurrency of training and integration with interruptions caused by problems, corrections, and workarounds affected the quality of the training experience. Both programs wrestled with aggressive schedules and the dilemma of when best to train the operators, considering the immaturity in the integrated product. A premature start elevates user frustration and provides an experience unlike actual operations, whereas a delayed start jeopardizes readiness for the operational mission.

In both cases the problems and changes during integration contributed to more training time than was allotted. Nevertheless, even with a stable, more mature integrated product, SOS mastery requires stepwise, incremental training.

In retrospect, the DPS training program provided too segmented a view of the SOS. The program succeeded admirably at training the workforce on individual systems, consistent with production assignments. But it allowed inadequate time in the integrated SOS environment, and the insufficiency was further exacerbated by changes and latent defects.

Although operator access to the DPS in E&R was an acknowledged objective, the availability was reduced because of integration schedule pressures and the Agency's decision to accelerate the production use of the DPS. Furthermore, all operators were not provided entrée to the SOS. Some of the training events at a single Production Center offered that access; most events combined subsets of the SOS integrated with interface simulators or partial configurations of hardware, or software modified for training.

All these limitations created a steep learning curve for operators to capitalize on the SOS and to adapt processes with it. The DPS production community was expected to respond to changing scenarios and requirements, essential for crisis response, which often necessitated actions different from normal production activities. Without the longer training investment on the fully integrated DPS, the production community struggled to develop alternatives to the new processes on which they had been trained. They frequently turned to non-DPS production systems and established processes.[2]

This situation was aggravated by the complex interrelationships of the various systems of the DPS and the limited information about them provided in the training. Over time this situation was mitigated by on-the-job experience and management efforts. A kind of "SOS team" approach later was introduced as a result of user dissatisfaction with the segmented view. But the experience underscores the need for training that provides a coherent understanding of the SOS.

An analogous result occurred with the TF XXI case. In assessing the force-on-force encounters at the NTC, participants and observers noted that when the EXFOR

---

[2] Numerous defects and changes in customer requirements exacerbated the frustration with the DPS.

engaged the OPFOR, it did not always exploit the new capabilities. Naylor (1997) observed:

> *...although the EXFOR troops have been learning about and training with the new systems, as well as developing tactics to exploit them, for more than a year, they are still not as adept with their new gear as conventionally equipped brigades are with current systems and doctrine. Thus, in many cases they failed to maximize their newfound capabilities.*

In fact, in many observed examples, the soldiers reverted to more conventional strategies and processes they knew better.

When later reconstructing the events at the NTC using the data collected during the experiment, the potential for capitalizing on the new digitization capabilities was evident, although sometimes in contrast to how events actually had transpired. This was why some participants characterized the training as the biggest shortfall.

The conclusion from this is that the EXFOR required more time and more iterations of experiences with the full complement of capabilities to evolve dramatically new strategies.

The post-NTC assessments underscored this conclusion. Effectiveness increased when several missions of the TF XXI AWE were replayed by the EXFOR and other participants using simulations. Repeated opportunities reinforced this result:

> *The consensus from those involved in analyzing the AWE was that numerous opportunities to exploit digitization were provided to the EXFOR, but they either failed to take advantage or failed to recognize their presence. Modeling three of these opportunities with different tools, at different locations, with different players led to a consistent finding that, if in fact the EXFOR had reacted to the information available in a timely fashion, then significant value-added to force effectiveness would have been observed. As with any new concept or system, knowing how to best employ it takes trial, error, and retrial.* (TRADOC Integrated Report Annotated Briefing Slides, p.57)

A Lesson About Training Operators

These two sets of experiences on training operators to use a SOS lead to a first lesson.

Training Lesson 1

Train operators on a SOS using a full spectrum of operational activities, and train allowing iterations.

Training should include iterative exercises with a full spectrum of operational activities using the SOS. This is in addition to the training provided on individual systems that are standalone or not integrated into the SOS.

Training on an individual system will not supply the necessary understanding of how the whole SOS functions or behaves. Such limitations in the training experience can inhibit operator effectiveness from two perspectives—capitalizing on the full benefits of the integrated whole and dealing with the relationships of other systems in the SOS.

Operating with SOS capabilities requires *more* training than operating with an individual system. The case studies show that increased time and types of training in the SOS environment were necessary to master the new capabilities offered by the integrated whole. This aspect of learning is not one-time, but rather iterative.

Both sets of experiences indicate the importance of training to assimilate the total picture of how the SOS supports the mission and how the individual systems contribute to the whole. The user must master his or her role but in the full context of the whole. For using a SOS capability, more than usual attention should be focused in training content on the interfaces among the constituent systems and how their relationships affect the operator. In actual operations, an operator must be prepared to respond when actions are initiated by other operators on other systems. While it may be difficult to anticipate the full spectrum of activities that a particular operator will experience with a SOS, the case studies reveal that the more opportunities the operator has, the better prepared he or she will be.

SOS training should use a wide spectrum of operational missions and a full complement of systems. The DMA experience illustrated the disadvantage of excluding some MC&G products from the suite of training exercises. Later, the operational community struggled to complete the detailed processes for producing the unexercised products and operators uncovered defects not previously manifested. Both training programs also reinforced the desirability of using a full complement of all constituent systems and equipment with little-to-no reliance on synthetic stimuli. Training with the SOS capabilities as near as possible to actual operational configurations enhanced the operator's preparation for production assignments.

This lesson is summarized neatly in a synopsis of the EXFOR's experience at the CTSF (Boutelle & Grasso, 1998):

> *The final ingredient in fielding a highly integrated set of capabilities is to ensure that the end user understands the power behind the system. The end user must make the system part of his or her business. Consequently training must not be focused on how to punch the keys, but on how to better conduct business. Training must also include the collective set of capabilities available to the end user.*

One TF XXI leader-soldier eloquently characterized the objective as follows:

> *I call this the fourth dimension of leadership—the full understanding of what the information technology is and what it can actually do for you. The new capabilities require a different*

thinking, an assimilation of the potential and the ability to leverage it faster. This requires new courses of action by iterations on exercise scenarios. Using these approaches, and by creating new and harder constraints on the operation, the leader will find ways to accomplish the end game. We must find ways to train for this holistic approach.

A Lesson on Infrastructure Support for Training

The case studies point to the need to supplement the training infrastructure to support training with a SOS, as noted in this next lesson. This augmentation is necessary even when the integration event is not as dynamic as was the case with both programs. The operator requires a training experience which provides the perspective of the SOS as a whole, and this carries implications for the training infrastructure—the people, processes, organization, components, materials, documentation, and automated implementation.

Training should not provide the perspective of a single discrete system absent the context of the whole. If training gives the operator a perspective primarily or exclusively that of the individual system, the training experience will naturally be less comprehensive, even inadequate, in exposing the relationships between and among the other capabilities in the SOS—or in providing the holistic view. This is a shortfall more readily caused when the constituents of the SOS are themselves existing systems; however the DPS though a "new start" program also provided too-segmented a perspective to the operators.

There will be many SOSs; systems will become constituents in many SOSs. Therefore there is a need for generation of new training content to teach the capabilities of each SOS, and to expose the new and different relationships. This will be a recurring need.

Training Lesson 2

The training infrastructure should be augmented to provide the perspective of a SOS.

The infrastructure supporting training should address the SOS supporting the mission, as well as the capabilities of the constituents and their relationships. Because operating with a SOS requires more training than operating with an individual system, an augmentation in the training infrastructure naturally follows.

The training infrastructure should be sufficiently robust to accommodate an iterative learning experience and training that is concurrent with integration. This requires a ready revision to training materials. Training content must advance as operators progressively master the capabilities of the whole. This need for flexibility carries special challenges when training content is *embedded* in individual systems.

Both programs felt the disadvantage of dealing with problems while SOS integration and operator training were concurrent. Despite this, cartographers and soldiers demonstrated great ingenuity and commitment in coping with the problems—and in getting their jobs accomplished. The Army leadership commented on the impact of focusing on integration to the detriment of training at the completion of the EXFOR's encounter with the OPFOR at the NTC (Naylor 1997). Also subsequent evaluations of results marked the adverse effects of immature systems on the EXFOR's training (TRADOC Integrated Report Annotated Briefing Slides, 1997; Slabodkin, 1997).

As a practical matter, training while integrating may be more representative of pressing operational missions. What then follows is the need to lessen the disadvantages when they are concurrent—or when there is a great rate of change in the SOS.[3] And there clearly are benefits to the integration processes when operators train while the integration continues. First, the nature and number of operator interactions increases the breadth of testing, improving the quality of the deployed SOS. Second, the intense experience by users can lead to an alignment of the integrated product with operational needs—when developers stand ready to respond. The TF XXI demonstrated this.

Training with immature systems and components and/ or a stabilizing integrated product will always be burdensome. When training is concurrent with integration more engineering support is required to respond to changes and defects to minimize

[3] In fact, the difficulty of training soldiers on new equipment with a high rate of software changes has been noted in prior exercises such as Focused Dispatch. The Army continues to evolve training methods for the digitized forces (Meigs, 1998; Ruocco & Smith, 1998).

interruptions and workarounds for the operator. Different operators will uncover different defects, even in a well exercised and stable SOS environment. This is characteristic of highly interactive information systems. And the Heraclitan principle[4] argues that the external environment, not just the one internal to the SOS, will cause changes to occur. The training support must be strengthened accordingly.

Impacts can also be lessened by refreshing training software and training data with the current SOS baselines and by updating training materials frequently. The timely dissemination of revisions to trainees is equally important. Without this added investment for support, the difference in the training experience from that of actual operations will be amplified. This disparity can detract from operators' performance. The anomalies and interruptions certainly lower user confidence and increase user frustration with the new capabilities, as both programs experienced.

More refresher training should be provided. There is a need to continue to educate users on changed capabilities. There is also a requirement to update a user's knowledge of the whole, as it is mastered, building upon earlier information for subsequent use.

Accommodating change while maintaining the currency of the training experience carries implications for planning and supporting the infrastructure. For the DPS program, the training infrastructure included the hardware and software baselines of the SOS, step-by-

---

[4] *"You can never experience the same SOS twice"* (see chap. 1).

step operational procedures, and data synchronized to those steps. All of these were resource-intensive to maintain when the software baselines of the individual systems were changing more rapidly than those used for the training materials. To accommodate changes and document workarounds for problems, considerable resources were used to annotate changes on training documentation and to distribute information to operators in training. Because the SOS software used for training was not updated as frequently as the SOS software used for integration, anomalies of experience and information resulted. Reconciliation was burdensome for the operators when they moved from training to a production assignment. Maintaining currency was complicated further by the different training configurations required for the three Production Centers. Analogously, the TF XXI faced equivalent demands to maintain the currency of the training experience for operators in light of substantial software revisions during integration. The program also experienced the challenges of training on nearly 200 configurations in the systems architecture.

If the training site is the integration site, as it was for the TF XXI at Fort Hood and for the DPS at a DMA Production Center, this infrastructure support for training becomes an important element of the environment there.

A Lesson About Training the Wizard's Team

The training programs for the DPS and the TF XXI were focused appropriately on preparing the operational community to accomplish the mission. There was little attention on the training needs of the players **behind the Wizard's curtain**. These included engineers, integrators, hardware and software developers, testers, infrastructure and administrative support personnel, and government and contractor managers. An implicit assumption was made in both programs that these participants understood the SOS.

Unlike the TF XXI experiment, the DPS was a production capability with a life-cycle that included maintenance and evolution; consequently, the DPS training program did address DMA personnel who had hardware and software maintenance responsibilities or who had the administration of the networks and the data bases. As it had for the DMA operator community, the training focused on individual systems, but was limited in addressing the interfaces between systems. This resulted in gaps in understanding that had to be overcome through on-the-job experience.

In light of the integration experiences,[5] these omissions result in a third lesson on training.

Training Lesson 3

> **Training should be provided to those *behind the Wizard's curtain.***

[5] See chap. 4.

Training people **behind the Wizard's curtain** about the SOS architecture being integrated contributes to their effectiveness and that of the process. Here the content of training provides information about the operational, technical, and systems architectures, and the relationships among the constituent systems of the specific SOS being integrated.

Both programs experienced shortfalls in personnel with the necessary knowledge of the SOS at the time of the integration. While their numbers were addressed rapidly through augmentation of assets, the transmission of knowledge was not. No training program was available to describe the SOS architecture and relationships among the individual systems with explicit "as-built" details in the context of the end-to-end mission scenarios operators were using.

The SOS team needed extensive knowledge to master the holistic view, and it was in short supply when it was needed most. The complex and detailed information was difficult to absorb as an on-the-job experience and required time to assimilate. For the DPS, with the relatively small size of the SOS cadre, only a small percentage of personnel actually achieved such mastery of the whole entity and the relationships among the constituent systems—about 1 percent of the participants. New personnel, particularly those rushed to the integration site, had to be educated, but there were only briefing materials compiled as a tutorial to use and they were more general, rather than detailed, in nature.

In hindsight, it was a training need that was overlooked and underestimated. With the large number[6] of people who supported both the integration phase and the subsequent sustainment, a training program will realize efficiencies and effectiveness.

In the context of rapidly configuring a SOS for any mission, jump-starting the knowledge and understanding of the Wizard's team must be considered as essential. A modicum of training also will provide universal understanding of the common processes, practices, and tools used in the SOS integration environment.

While it is of primary importance that the SOS cadre managing integration receive such training, the experiences show that the teaming of all participants **behind the Wizard's curtain** was critical to success. The benefits of collaboration and the synergy from the diverse and specific views of the individual systems in the SOS argues for providing training to a broader population, which includes personnel supporting the constituent systems.

Training the participants for the integration of a SOS and for evolution and maintenance requires resources. Sustaining current training materials takes increased effort, just as it does to train the operational community to support the mission. The complex and dynamic nature of the SOS capability presents special challenges for education and presentation.

[6] See discussion in chap. 5, lesson 4.

Chapter 7

An Integration Environment for the Future

Considerable efforts were required to integrate the DPS and the TF XXI. Yet each was a relatively simple system of systems (SOS) in the context of the Joint Vision 2010 era. Operations typically will be joint and coalition. It is important to look toward integrations that go beyond the complexity of the two case studies, and consider federations of systems (FOS) as well. This chapter briefly explores these topics and concludes that future experimentation should include assessments of SOS and FOS activities **behind the Wizard's curtain**.

Nonetheless, the lessons from both programs provide a foundation upon which to build. Several recommendations are given in this chapter, including the use of an integration environment. This approach incorporates and supplements that of the U.S. Army's Central Technical Support Facility (CTSF), so essential to the TF XXI experiment. For integrating a FOS, further refinements will be needed.

Some changes in the acquisition culture are recommended as well. A template for future work is put forward—many open questions remain to be resolved. Attention should be directed to the entire life cycle of a SOS and FOS, whereas this book has focused on the integration phase. Finally, the link between an integration environment and future training is discussed.

Recent Coalition Operations

Coalition operations are intrinsic to the era of Joint Vision 2010. Yet problems with even rudimentary interoperability occur today in coalition operations—as demonstrated in *Operation Restore Hope* (Starr, 1996). The compilation of lessons from *Operation Joint Endeavor* in Bosnia by Larry Wentz (1998) illustrates the difficulty. So detailed is his examination that it comprises the equivalent of a case study, one focused on a recent coalition operation.

The environments where today's forces must function are increasingly complex and heterogeneous and will grow more so. While U.S. defense systems are managed by separate organizations in consonance with a defense enterprise architecture, coalition partners bring their own information systems. These are and will be developed with different methodology, technology, and standards. There will be greater diversity along with multiple architectural frameworks, as was the situation in Bosnia.

According to Wentz (p 273):

> *Coalition operations such as Joint Endeavor present a complex set of challenges for the*

> *military C4ISR[1] system planners, implementers, and operators. The most difficult challenge is the provision of integrated C4ISR services and capabilities to support the needs of ad hoc multinational military force structures and politically driven command arrangements. Although integrated C4ISR services are the desired objective, the realities tend to drive the solution to stove-piped implementations. In spite of technology advances, this will likely be the case for some time to come. There will continue to be uneven C4ISR capabilities among coalition members who will continue to rely on systems with which they are most comfortable—their own.*

Wentz observes that this situation in Bosnia resulted partially from the widely disparate information technology capabilities in and among the coalition memberships. Often these were used in the absence of in-country infrastructure. Examples abounded of infrastructures comprised of diverse parts. There were independent and separately managed NATO systems for voice, messages, and data and video networks. The national tactical assets of the framework nations provided telecommunication capabilities because the Bosnia infrastructure was destroyed, some of it by NATO air strikes. These were supplemented with United Nations satellite terminals and commercial products and services.

[1] Command, Control, Communications, Computers, Intelligence, Surveillance, and Reconnaissance.

There was a strategy to use commercial products and services, but interoperability between military and commercial systems continued to plague forces. Numerous examples occurred where strategic, theater, and tactical systems had difficulties in exchanging information. Networking applications also were not interoperable. As anticipated, language was an interoperability issue with 34 separate nations participating (Wentz, pp. 351-353).

Various systems were acquired by many different coalition organizations, such as NATO. Some required integration with the information systems of non-government organizations and with those of private volunteer organizations. In Bosnia this situation was complicated by the number of civilian and non-government organizations with whom information had to be exchanged, and with whom activities required coordination. For example, civil activities collectively constituted an important intelligence asset. Some non-governmental organizations, private volunteer organizations, and international organizations were included in intelligence collection plans. Often such organizations relied on information infrastructures separate from those of the government or military organizations.

While Wentz asserts problems will continue, he acknowledges progress toward interoperability and credits the good efforts of people in various organizations who did achieve a successful integration of the disparate systems. However, the achievement consumed *time and resources*. To integrate communications and information systems, it required 9 months of planning and lead roles by three nations to establish the capability to allow the Implementation Force (IFOR) to take command and control of the operation (Wentz, pp. 280-282).

The Joint Vision 2010 Era and Federations of Systems

The recent experiences in Bosnia presage the future era. While students of Joint Vision 2010 acknowledge its reliance on a SOS, the operational needs for many missions will require the capability of a FOS[2] as well, where management of development and operations is collaborative. Joint Vision 2010 embodies certain operational concepts that rely on collaboration rather than central direction and management. However, how these are actually implemented is still to be determined.

Joint Vision 2010 relies on partnerships, as typified by the situation in Bosnia. As in Bosnia, systems in a FOS are acquired and managed by many organizations, including foreign, civil, and commercial organizations, military and non-military, governmental and non-governmental. Usually all partnerships are not the same; relationships and information sharing differ. In some operations, there is even ad hoc participation by organizations (and their systems), as also occurred in Somalia during *Operation Restore Hope*.

---

[2] S/FOS characteristics are discussed in chap. 2.

With this reliance comes the heterogeneity of the FOS. This derives from the systems architectures that are the legacies of many nations and many organizations. Diversity will propagate in the architectures through the application of different technical standards and through the rich mix of multiple cultures, languages, and semantics used to interpret them. The FOS also mirrors the diversity in the differing relationships of the partners and their various operational architectures. The coalition organizations use systems that are aligned with their individual requirements, organizational structures, and operating scenarios.

In the Joint Vision 2010 era, the local commander will be given more assets, more information, and more control over them. There will be more management control exercised at the tactical level, and in some cases to the level of the individual combatant. This is a concept more consistent with federation principles where power is delegated, usually to the lowest possible unit in an organization.

However these U.S. operational scenarios may be substantially different from those of other coalition partners. Not only are organizational structures of international partners frequently different, so are their approaches to command and the rate at which decisions are made. In a coalition mission, control may be accomplished at quite different levels by the various participants. These differences in their operational scenarios migrate into implicit differences in their information systems, giving rise to a whole body of effects that become obvious only when they are exercised together.

There are many relationships and interdependencies that will increase in complexity in the Joint Vision 2010 era. Maneuverability by forces will require not only greater dispersion but also more agility across global distances. The use of coalition assets is critical to attain global reach. The global dispersion of joint capabilities and assets—at strategic, theater, and tactical levels—must collaborate in execution for massed effects.

The consequence of operating with greater dependency on collaboration, increased heterogeneity, and greatly dispersed assets is increased vulnerability to failures of interoperability and integration. The implications of a FOS are portentous. Failure to achieve an integrated product could bring serious consequences to an operational mission or result in capabilities that fail to provide the best operational advantage.

A Way Ahead

The more complex future undermines the viability of integrating the disparate parts of a FOS or a SOS, or integrating them with limitations on time and resources. Certain strategies adapted in both the DPS and TF XXI programs proved beneficial. For now, these provide a foundation for a future SOS or a FOS, but more work is necessary. Several recommendations, including a template for future work, are summarized in figure 7–1.

A Way Ahead

Direct attention *behind the Wizard's curtain* in future experiments

- ✧ Include coalition partners in experiments

Use an integration environment

Evaluate more case studies

Use a single facility for integration

- ✧ Investigate distributed integrations

Develop acquisition specialists for SOS and FOS

Address best practices for the life cycle

Tailor methods for estimating time and effort

- ✧ Compile SOS project databases

Address methods to train for a SOS and FOS

Figure 7-1. A Way Ahead

Direct Attention **Behind the Wizard's Curtain**

There will be opportunities in the future to scrutinize, assess, and improve activities **behind the Wizard's curtain**—if the attention is directed. The implementation of Joint Vision 2010 will proceed from concepts to joint warfighting capabilities through a long-term continuous process. (Shelton,1997–1998, 1998; Reimer, 1997–1998; Coats, 1997–1998; Hallion, 1997–1998; Barnett, 1997–1998; Hoffman, 1997–1998). Joint and service experimentation, exercises, and demonstrations will continue to comprise elements of the evolution. The commitment to experimentation as a means to shape the implementation of Joint Vision 2010 has led to the designation of the U.S. Atlantic Command as the leader for joint experimentation. As activities proceed, various SOS or FOS (S/FOS) will be configured; consequently, attention can be focused on the efforts **behind the Wizard's curtain** in addition to that directed on the operational performances playing "front and center stage."

And to address the realities of the future, experimentation should include coalition partners engaged in a spectrum of operations using their own systems integrated into a S/FOS.

Use an Integration Environment

To integrate a S/FOS, an integration environment should be used. Here this environment is labeled SFIE.[3] It is a concept, not an organization. It is the environment

[3] S/FOS integration environment.

of people, processes, and infrastructure used by a team consisting of acquisition and operational personnel to manage the integration before the product is deployed for an operation or an experiment and to sustain it afterward. As such, it describes who and what appear **behind the Wizard's curtain**. The environment can be convened by any organization and constituted in any appropriate facility. The SFIE is a mechanism to achieve the results required, to accelerate the integration process, and to garner sufficient quality in the product. Otherwise, the operational forces must expend the effort in the field—absent the critical mass of engineering expertise. Not only do the two case studies demonstrate this need when revolutionary capabilities are involved, but a similar lesson was derived in *Operation Joint Endeavor*:

> *Exercises and training demonstrated the value of setting up the expected C4I configurations in advance of the deployment to sort out integration and interoperability problems. The exercises also served to train and do some team building for those personnel who would deploy.* (Wentz, p. 376)

Use of an integration environment does not replace the need to acquire and test individual systems and certify their compliance to the overall architecture of the SOS.[4] Nor does it replace the earlier phases of the life cycle. As methodology, the efforts in the integration environment are *additive* to the essential activities

[4] See the discussion on Lesson 1 in chap. 5.

accomplished before the integration begins, not "in lieu of." Using a SFIE does not negate any efforts by the services and agencies to develop and acquire robust system capabilities and build common architectures. Rather the application of the SFIE is tractable with the continuation of such efforts.

Elements Included in the Integration Environment

The case studies indicate the need for a set of processes and infrastructure that are common for all teams participating in the integration—and established at the integration facility. This set is fundamental and intended as minimal—because the constituent systems of a S/FOS are managed, developed, and operated by many different organizations for their own purposes with their own processes and infrastructure.

The elements of the SFIE, compiled directly from the lessons of the case studies, are listed in Appendix B. The SFIE includes leadership empowered for the integration, a S/FOS cadre, and participants who manage, develop, and operate the individual S/FOS systems.

The inclusion of a "core" in the common infrastructure is necessary for integration[5] and sustainment of the S/FOS during mission operation—to accommodate changes, such as from technology insertion, COTS turnover, and corrections. While this

[5] A core is the minimum set of all the hardware components, all the software, and the architectural framework in the S/FOS. See chap. 5, lesson 3.

constitutes something of an investment, its value must be assessed against the risk of unanticipated and adverse impacts from changes while the integrated product is deployed in the field. For the DPS core, the investment was less than 1 percent of the cost of the total program.

This expenditure can provide added advantages for purposes of training—but only with augmentation if the training population is large.[6] Access to the S/FOS can facilitate the iterative learning process needed to master the capabilities of the integrated product, while providing training on the integrated capability actually deployed. These were important benefits experienced in the two case studies and in *Operation Joint Endeavor*.

The Need for More Case Studies

Two case studies were used. Therefore it cannot be argued that these common processes and infrastructure in the SFIE constitute the *complete set* needed. While it was tempting to include more, caution is warranted when the learning curve by participants could be steep, when adaptation by individual teams consumes schedule and resources, and when technological implementation of the common infrastructure requires continuous change.

More assessments are required to determine other elements that should be standardized and those that can

[6] For verification, a core may require only a single workstation of a unique type. Hundreds of workstations would be required to train hundreds of operators concurrently.

remain particularized to the individual teams. In reviewing both case studies, individual teams retained many processes and tools that remained uniquely their own. For example, none of the individual teams participating in the DPS or TF XXI had identical quality assurance processes or tools.

Additional case studies would provide the means to compare whether methods for one SOS unequivocally apply to another and to a FOS as well. For example, determining the sensitivities of approaches to numbers of systems in the set or to measures of coupling[7] between systems, or to the degree of autonomy, heterogeneity, and dispersion are comparative analyses of interest.

Test environments requiring integration of multiple systems have been applied by the services. One example is that of the Fort Franklin/CUBE,[8] used by the U.S. Air Force to manage the problems of technology insertion during *Operation Joint Endeavor* (Starr, 1996; Wentz). This also illustrated the advantages of retaining a core SOS to sustain an integrated product as changes were required for a continuing operation.

[7] Two systems are coupled if they are interdependent (i.e., if at least one system requires information from the other, or requires components, services, or people).

[8] Technology insertion was a major problem in Operation Joint Endeavor. Wentz reports on replication of C3I systems for the CAOC in Vicenza at a laboratory called the C2 Unified Battlespace Environment (CUBE), at Air Force Electronic Systems Center at Hanscom AFB. This was used for system integration testing of new capabilities before operational deployment to theater (p. 366-368).

Another rich source for examination is the effort to address the Year 2000 defect, conducted on a global scale. Individual systems are being tested, certified for compliance, and reintegrated into enterprises of systems. The successes (and failures) of the methods used will render important lessons, many of which will apply to a S/FOS.

Periodic evaluations of the dynamics of integration in a S/FOS are warranted, given the march to a more enterprise-centric strategy in the Department of Defense, discussed in chapter 1. Neither case study used in this work was able to take advantage of the (now available) joint technical architecture,[9] and more robust defense information infrastructure and common operating environment—because of timing and phasing. The Army's TF XXI established a common operating environment using surrogates for the defense enterprise applications, many of them commercial products. Much of the first version of the joint technical architecture was based on the Army's technical architecture. How these new(er) versions of the defense enterprise might change the lessons from future SOS and FOS integrations should be evaluated.

More studies should assess integration environments with coalition partners. The differing methods used by the various teams engaged in a FOS integration coupled with their multiple cultures and languages present challenges to defining common processes and infrastructure. Achieving collaboration among

[9] The Army's technical architecture was a primary source for the defense enterprise Joint Technical Architecture (JTA).

diverse communities of participants may require different strategies and incentives. There are examples of SOSs and FOSs emerging from the private sector (Maier, 1998) that also provide additional material for evaluation.

A Word on Leadership

The two case studies provide limited insight for managing **behind the Wizard's curtain** when the environment is that of a FOS, which is collaborative in nature. Rather, both programs exercised some form of direction over the various teams participating. Both examples illustrate the advantage of someone in charge. The overt empowerment of leadership of the integrations was a strategy successfully used.

The two programs were simple in organizational structure. They were managed by a single agency and a single service, respectively. The responsibility, accountability, and resources resided within a single organization for each.

The compelling need in future S/FOS ventures when it is not clear "who is in charge" is to resolve that ambiguity at the outset—or to determine a process for resolving it.

The leadership and management structures best suited for integration environments which are collaborative or voluntary and ad hoc, rather than directed in nature, should evolve. Recent coalition operations and future Joint Vision 2010 experiments with coalition partners should be assessed. Protecting the U.S. critical infrastructure, the subject of national attention, requires collaboration among federal, state, local, and non-governmental organizations; the resulting management structures may provide examples which are applicable (Presidential Commission Report, 1997).

Use a Single Facility

The lessons from the two case studies concluded that the use of a single facility with a supporting environment of common processes and infrastructure was essential to the success of their integration events. The management of each program made early use of geographically dispersed and distributed integrations—of individual systems, and of subsets of the SOS. However, to achieve the integrated product before deployment, each installed the set of all the individual systems that comprised the SOS at a single facility. And only limited use was made of substitutes for individual systems and components.

In an era of information technology and virtual collaboration, it is perhaps surprising to express the need for one real physical site for integration rather than for a virtual facility, but the evidence of both cases argues strongly[10] for this. With the importance of managing not only the achievement of a complex integrated capability from diverse parts, but also in constructing one team from many, the collocation of teams can contribute mightily. And for a military mission, the

[10] See chap. 5, lesson 5.

operational community may be placed at increased risk if the integration is insufficient.

The management of a FOS integration is more challenging than that of a SOS because collaborative relationships must be developed and sustained across a broader spectrum of heterogeneous cultures, organizations, developments, and operations. The single facility and common environment can bring cohesiveness to build the twin citizenship necessary for federations to succeed (Handy, 1992).

A facility with an integration environment could be designated when a mission is initiated. A S/FOS cadre could be assembled temporarily with members drawn from various organizations and countries. However, creating at least some level of permanent staff and facility is also an option (e.g., the Army has done this for integration with its CTSF at Fort Hood). This approach has the merit of building resident expertise, expediting establishment of the integration environment, and accelerating the integration processes.

Investigate Distributed Integrations

For coalition or joint missions, one integration site may not be feasible. The question that must be addressed is: if collocation is not viable, what environments (if any), virtual and distributed, can lend the same success and effectiveness—to support the integration of all systems in the set? Virtual environments must expand well beyond tools such as video teleconferencing (extensively used by both programs) to attain the communications, insights, and synergies required for success.

Future experimentation must establish if collaborating but dispersed service integration environments are sufficient for an integration of a S/FOS that supports joint missions—and sufficient for an integration of a S/FOS that supports joint missions and sufficient for integration of a FOS which supports coalition missions. The method for distributing the integration may hold the key to effectiveness. Partitioning subsets of the S/FOS by mission area (command and control) is one approach. There may be effective distributions determinable when entire end-to-end threads of activities to support a mission are exercised on the S/FOS by specific (but different) operational communities.

If integration proceeds incrementally with ever-growing numbers of systems and subsets using distributed environments, can the sufficiency of the integration process for all systems before deployment be preserved? This is a question that should not be too quickly answered or dismissed, but the evidence of the case studies suggests that "no" is the current answer. And the reasons extend beyond just the integration of constituent systems to the cohesion of the operational viewpoints and of the Wizard's team.

Develop Acquisition Specialists

The Heraclitan principle says, *"You can never experience the same SOS twice."* The future will require many integrations of a SOS or a FOS. It is important to have an appropriately skilled and sufficiently staffed cadre of people responsible.

A S/FOS cadre should be developed within the acquisition community. Because it is the whole, rather than the parts, that delivers the required results the future U.S. defense strategy is based on, the acquisition culture should endorse the merit and evolve the skills for producing the integrated product. The expertise required is in short supply, even in the private sector, and must be developed, as well as retained, in sufficient numbers.

Such roles are distinct from those of the program managers of individual systems. Program managers will continue to develop and maintain individual and independent systems to serve other and various missions. Within the acquisition environment, they will continue to have considerable authority, control significant dollars, and consequently be provided significant recognition.

A S/FOS cadre merits equivalent recognition and position. In addition to providing needed expertise, such specialists could evolve and reinforce architectures that are enterprise-centric for joint and coalition missions, while instantiating processes for integrating independently managed constituents. As missions cross institutional lines, the program and engineering adjudication process intrinsic to the cadre transcends the interests of parent organizations as well as of individual systems. Institutionalizing the cadre could promote universal recognition as well as provide a necessary balance in authority to that of the program managers who must manage within their own cost, schedule, and performance constraints.

The DPS case study illustrated how an Agency's culture affected the SOS cadre. The acquisition teams of the individual systems delivered functional capabilities identified with *core business* of the Agency. They were more easily recognized as contributing to Agency business than those in the SOS cadre. Less value and responsibility were perceived in engineering a SOS than in delivering a production system. Furthermore the SOS cadre was considered necessary only until FOC, when it was reduced in number and rank. Unfortunately this occurred just when the need to evolve the DPS became greater.

Life Cycle of System of Systems and Federation of Systems

While this book has focused on the integration phase, attention to all the phases of the life cycle of a S/FOS is warranted. Both program ventures dealt with challenges other than those of integration. For example, the architecting processes for the TF XXI allocated requirements across a broad spectrum of existing and developmental systems and their interfaces. This consumed time, requiring 7 months for an initial architectural baseline, which subsequently was refined.

For other ventures, the best strategies to expedite and optimize this process would be advantageous.

A guidebook for managers should be developed for a S/FOS; architectural and design principles tailored for a S/FOS would provide substantial benefits. These could be derived from those practices and lessons compiled from appropriate case studies and program ventures. An example of an analogous reference that delineates best practices and supporting tools is the DoD Software Acquisition Best Practices Initiative (1997), and there are others available.[11] However, as are many, it is oriented to the management of software projects.

Considerable work is proceeding to frame the practices and processes for software-intensive projects and those for large and complex information systems into an integrated whole (Schaeffer, 1998). There are current and emerging standards for engineering a system and addressing the life-cycle processes; these are discussed in Wright (1998). An assessment of practices and processes specifically tailored for a S/FOS and evolved from this integrated framework merits attention.

Address Architecting in Best Practices

A key question is: How does a manager develop the architecture for a S/FOS? While the U.S. defense enterprise provides a substantial framework of common systems and applications as a foundation, there will be

[11] The Defense Systems Management College disseminates information on best practices and lessons learned for the acquisition work force, available through its web site (Defense Systems Management College).

other and various systems in the set of systems used for a specific mission. And in dealing with a FOS, there is greater heterogeneity. This brings increased problems in interoperability, and therefore in integration.

The diversity in architectural frameworks in the FOS and the plethora of legacy systems contributed by the participating organizations compounds the challenge. Different standards and guidelines not only can impose different rules, but de facto the sets of rules may not be interoperable. Perhaps more problematic, because it is less obvious, are the more subtle architectural disconnects that arise from systems designed for different purposes or by different development communities operating with implicit understandings of their standards and guidelines that are not readily assimilated, traceable, or recoverable. And this can and does occur even with commercial products and services.[12]

One approach lies in reducing the multiplicity of frameworks, possibly by narrowing the choices of standards, similar to the mandate of the joint technical architecture within the U.S. defense enterprise. At least one recommendation has been made to accomplish this through the application of commercial standards (Commission on Physical Sciences, Mathematics, and Applications, "Realizing the Potential of C4I: Fundamental Challenges," 1999). Perhaps a coalition

[12] One of the effects of increasing reliance on COTS packages may be the introduction of architectural mismatches because commercial products are developed for various architectural frameworks (Gacek & Boehm, 1998).

technical architecture is viable, but it will not occur quickly, but rather evolve deliberately, rigorously, and (probably) incrementally with specific coalition partners. Another strategy is to allow the multiplicity but use automated means to identify and reconcile architectural disconnects or to translate diversity into commonality. Even if feasible, this also is not an optimum or short-term strategy.

Adapting architecting principles for a S/FOS in the context of Joint Vision 2010 would address perhaps the most important phase of the acquisition life cycle. However there is more basic work required, including general agreement on taxonomy and those characteristics which should be used to distinguish classes of "systems of systems." (In this book the degree of autonomy, heterogeneity, and dispersion have been used.) Bounding a SOS in an enterprise of many systems or bounding a FOS in the enterprises of coalition partners also present fundamental challenges, the strategies for which are not currently obvious.

In addition to a proposed taxonomy, Maier (1998) already has provided some basic principles for architecting systems of systems through several heuristics that are refinements of more general guidelines. The Army's success in developing an architecture for the TF XXI SOS, using the viewpoints of operational, systems, and technical architecture, provides a good example of architecting processes. The fact that the technical architecture used for the experiment correlated highly to the initial version of the joint technical architecture makes the TF XXI example a significant one for practices related to the defense enterprise.

A compilation of principles and design guidelines would benefit from those used in prominent examples like the Internet, as well as from more robust applications that are emerging in various business sectors such as intelligent transport systems (Maier, 1997).

Refine Methods for Estimating Time and Effort

For conducting activities **behind the Wizard's curtain**, it is important to plan adequately. To do this requires the ability to estimate cost and schedule resources for producing a S/FOS from a collection of independently developed and operated systems. The lesson[13] derived from the two case studies points to the need to plan for significant time and effort. However, the discussion flagged difficulties with good estimation methods and models tailored to a S/FOS.

Typical models available are based on a history of software projects. Many models have evolved using experience-based estimation, based on project data bases compiled over the lives of numerous projects. Other models use parameters, but are tailored to specific domains, such as the military development environment. A good summary of estimation methodology is provided in "Software Estimating Technology" (Stutzke).

---

[13] See chap. 5, lesson 4.

Resources can be estimated using various factors that characterize the project, such as development approaches, product complexity, team experience, and work environments (Boehm, et al., 1996).They can be projected for different phases of the life cycle as well. However, even for conventional projects, Capers Jones (1998) notes that most tools available today are inadequate in treating data bases, multiple domains,[14] and enterprise estimation. And they are calibrated to projects different than a S/FOS.

A S/FOS comprises existing systems independently developed for other and specific purposes. Tailored estimation methods are needed for projects that include a single enterprise as well as multiple enterprises, and with components that include substantial numbers of legacy systems as well as developmental systems.

Good estimation processes also require a data base of projects that are analogous. Royce (1998, p 29) says:

> *Extrapolating from a good estimate, an ideal estimate would be derived from a mature cost model with an experience base that reflects multiple similar projects done by the same team with the same mature processes and tools.*

Two developments are needed for more accurate estimation of resources for a S/FOS—tailored methods and an empirical project data base. The opportunity to compile data on S/FOS activities **behind the Wizard's**

[14] Products constructed from hardware components, software components, data base components, and microcoded components.

curtain comes in conjunction with the Joint Vision 2010 experimentation and from other S/FOS projects. Compilation of data coupled with the best methods for estimation should evolve to enable estimation a priori, for each phase of the life cycle, and for each specific S/FOS.

Address Training with a System of Systems and Federation of Systems

During the integration phase of both case studies, the operational community trained using the integrated product during integration. The lessons learned [15] can be translated into three training principles applicable to missions that require a S/FOS:

- Train operators on the S/FOS in iterations
- Train operators for the S/FOS, not just for individual systems
- Train the team **behind the Wizard's curtain**.

Determining the best methods to implement these simple principles needs to be addressed.

During integration, changes (including corrections and adaptations) are continuous. They occur in the constituent systems and all the relationships. The simple approach of waiting to train until stability is reached is not an option—because the mission schedule is usually urgent, because the training population is large, and because Heraclitus is right. A window of time without

[15] See chap. 6, training lessons 1–3.

changes is a small one. Developing the best methods to offset the adverse effects of change during integration while training to teach the synergy of the whole using the capability deployed deserves future attention.

The dynamics of integration complicate the task of providing and maintaining a current perspective of the whole—which emerges more accurately with the integration and evolves through the operational use. The orientation of the operator to an individual system should be supplemented with the perspective of the S/FOS—then refreshed. And the orientation needs to be reformed for each mission, when a different SOS and FOS are required. The relationships within and the synergy of the whole are different.

The dynamics of integration complicate the training of the teams **behind the Wizard's curtain**. "Training" in this context is about the S/FOS architecture—the operational, technical, and systems viewpoints. Many participants bring perspectives appropriately but entirely focused on the management and development of their own systems. Providing a current and common understanding of the architecture and the relationships among the constituent systems while communicating the dynamic and emerging behavior of the whole entity is a training need that should not be overlooked. The methods dealing with change and maintaining current information on the integrated product for training the operational community are applicable for this population as well, but the means remain to be determined.

Realizing the Promise

The concluding recommendation is the same as the first—attention must be directed at those activities **behind the Wizard's curtain** that produce, as well as integrate, a SOS or FOS. The joint experimentation for Joint Vision 2010 provides the opportunity. The case studies and current coalition operations presage the complexity of future ventures. The foundation established through the case studies—the lessons learned, and the integration environment—are neither a final solution nor a silver bullet for the Joint Vision 2010 era. Future experimentation and future assessments are essential.

The U.S. defense strategy and Joint Vision 2010, as envisioned, rely on an integration of systems to form a SOS, and, implicitly, a FOS. Many different combinations of systems will be required based on the type of mission and the nature of the geopolitical environment. *"You can never experience the same SOS twice."* There is not one SOS or one FOS, but many. This reinforces the need to continue to examine the processes to determine how systems, independently managed, developed, and operated, may be integrated more quickly and more effectively.

There will be future opportunities to scrutinize all aspects of producing, integrating, deploying, and sustaining a SOS and a FOS. But the data must be compiled and the analyses conducted for the Wizard's team. Most importantly, as lessons are learned through these examinations, they must be used as a foundation for better principles and practices which address every aspect of a successful SOS and FOS. This evolution is necessary to produce the technological magic for realizing the promise of Joint Vision 2010. To continue the assessments of activities **behind the Wizard's curtain** is the most important message of this work.

Appendix A

The Lessons Learned

Nine Lessons Learned on Integration

Lesson 1

Certain activities should precede a SOS integration. These include:

defining the SOS architecture;

developing and testing the individual system constituents of the SOS;

developing and testing the interfaces between and among the individual systems of the SOS;

independently certifying compliance with the SOS architecture.

Lesson 2

Use early, incremental, and iterative integration to achieve a SOS.

Lesson 3

The testing strategy for the integration of a SOS requires:

an agreed-to plan and process for testing, based on a risk assessment;

a suite of activities representative of the operational requirements of the mission the SOS supports;

the exercising of a full spectrum of the SOS activities (end-to-end) by operators, using the actual constituent systems of the SOS—or at least a core SOS.

Lesson 4

To integrate all the systems of a SOS, plan for substantial difficulties and significant time and resources.

Lesson 5

The use of a single facility—with an environment of people, processes, and infrastructure—substantially facilitates the integration of a SOS from individual systems.

Lesson 6

The process for SOS integration should overtly address the leadership of the integration as follows:

an on-site acquisition leader empowered for the integration of the SOS and an on-site leader empowered for the operational community;

supported by a SOS cadre——with sufficient resources and authority;

supported by participants who manage, develop, and operate the constituent systems of the SOS.

Lesson 7

Certain common processes and common infrastructure in the integration environment are essential to manage a SOS integration successfully. These include the following:

an Engineering Board with responsibility and authority for identification and resolution of SOS issues and discrepancies, including the assignment of responsibility for correction;

establishment of processes (and the automated means) for identification of SOS issues and discrepancies, their disposition, tracking, and resolution, under the management of the Engineering Board;

automated support for the tracking and tracing of SOS operational requirements;

configuration management and control of the hardware and software baselines of the systems of the SOS by the integration leadership, supported with: automated means for identifying and controlling the baselines and subsequent changes; a formal build, verification, and re-integration process for changes;

a robust communications infrastructure linking the teams internal to the integration environment and their external counterparts;

an office automation environment to support the integration's administrative processes as well as to support interpersonal processing and communications for the participants.

Lesson 8

Certain common processes and infrastructure in the SOS integration environment promote effectiveness and efficiencies. These include:

daily planning and scheduling of resources (people, equipment, facilities) for integration events—with contingency plans and schedules readily available;

timely dissemination of information pertinent to each integration event, such as test status, equipment availability, and results;

daily status meetings, with results immediately available.

Lesson 9

Prototyping a SOS can provide early insight into operational requirements and into the SOS systems architecture.

Three Lessons Learned on Training

Training Lesson 1

Train operators on a SOS using a full spectrum of operational activities, and train allowing iterations.

Training Lesson 2

The training infrastructure should be augmented to provide the perspective of a SOS.

Training Lesson 3

Training should be provided to those *behind the Wizard's curtain.*

Appendix B

The Integration Environment

The elements of the integration environment for a SOS and a FOS (S/FOS) include:

An integration facility with its internal administrative infrastructure;

A team consisting of:

> **an on-site acquisition leader empowered for the integration of the S/FOS and an on-site operational leader empowered for the operational community;**
>
> **a S/FOS cadre;**
>
> **participants from organizations who manage, develop, and operate the constituent systems of the S/FOS.**

Common processes consisting of:

> **a testing strategy for integration which includes a suite of activities representative of the full spectrum of missions the S/FOS supports; exercising of those activities end-to-end by operators, on the actual or core S/FOS;**

an engineering board with the authority for identification and resolution of S/FOS issues and discrepancies, including the assignment of responsibility for correction;

the tracking and tracing of S/FOS operational requirements;

configuration management and control of the hardware and software baselines of the systems of the S/FOS, including a formal build, verification, and re-integration processes for changes;

daily planning and scheduling of activities, including contingencies;

the dissemination of information pertinent to each integration event;

the dissemination of daily status.

Common infrastructure consisting of:

the actual S/FOS or a core S/FOS;

a robust communications infrastructure linking the teams internal to the integration environment and their external counterparts;

an office automation infrastructure to support administrative processes of the integration.

Appendix C

Acronym List

4ID—Fourth Infantry Division

A2C2S—Army Airborne Command and Control System

ABCS—Army Battle Command System

ABIS—Advanced Battlespace Information System

ACM—Association for Computing Machinery

ACT—Activation Control Team

ADA—Air Defense Artillery

ADO—Army Digitization Office

AFATDS—Advanced Field Artillery Tactical Data System

AMC—U.S. Army Materiel Command

AMPS—Aviation Mission Planning System

AN/TPQ–36v8—Firefinder Radar System

ANBACIS—Automated Nuclear Biological and Chemical Information System

ASAS—All Source Analysis System

ATA—Army Technical Architecture

ATCCS—Army Tactical Command and Control System

ATM—Asynchronous Transfer Mode

ATMT—Agency Transition Management Team

AVN—Aviation

AVTOC—Aviation Tactical Operations Center

AWE—Advanced Warfighting Experiment

B2—Brigade and Below

BCIS—Battlefield Combat Identification System

BCV—Battle Command Vehicle

BDE—Brigade

BN—Battalion

BSFV-E—Bradley Stinger Fighting Vehicle–Enhanced

BTRY—Battery

C2—Command and Control

C2V—Command and Control Vehicle

C4I—Command, Control, Communications, Computers, and Intelligence

C4ISR—Command, Control, Communications, Computers, Intelligence, Surveillance, and Reconnaissance

CCRP—C4ISR Cooperative Research Program

CGSP—COTS Ground Station–Prototype

CIS—Communications and Information Systems

COE—Common Operating Environment

CSSCS—Combat Service Support Control System

CTSF—Central Technical Support Facility

CUBE—Command and Control Unified Battlespace Environment

DBS—Direct Broadcast Satellite

DII—Defense Information Infrastructure

DIL—Digital Integration Laboratory

DISC4—Director of Information Systems, Command, Control, Communications, and Computers

DMA—Defense Mapping Agency

DPS—Digital Production System

E&R—Exercises and Rehearsals

ENG—Engineer

EPLRS—Enhanced Position Location Reporting System

EXFOR—Experimental Force

FAAD—Forward Area Air Defense

FAASV—Field Artillery Ammunition Supply Vehicle

FOC—Full Operational Capability

FORSCOM—U.S. Army Forces Command

FOS—Federation of Systems

FSB—Forward Support Battalion

GBS—Ground Based Sensor

GCCS—Global Command and Control System

GPS—Global Positioning System

HH—Handheld

IEWCS—Integrated Electronic Warfare Common Sensor

IFOR—Implementation Force

INC—Internet Controller

IOC—Initial Operating Capability

IREMBASS—Improved Remotely Monitored Battlefield Sensor System

ISR—Intelligence, Surveillance, and Reconnaissance

JSTARS—Joint Surveillance Target Attack Radar System

JTA—Joint Technical Architecture

LAN—Local Area Network

LGSM—Light Ground Station Module

LINC—Lightweight Internet Controller

LLDR—Lightweight Laser Designator and Rangefinder

LRAS3—Long Range Advanced Scout Surveillance System

M1A1—M1 Main Battle Tank, Improved Version

MC&G—Mapping, Charting, and Geodesy

MCS/P—Maneuver Control System/Phoenix

MCS—Maneuver Control System

MECH—Mechanized

MET MEAS—Meteorological Measuring Set

MFCS—Mortar Fire Control System

MI—Military Intelligence

MP—Military Police

MSE—Mobile Subscriber Equipment

NIMA—National Imagery and Mapping Agency

NMT—Network Management Terminal

NTC—National Training Center

OH-58D—Kiowa Warrior Helicopter

OOTW—Operations Other Than War

OPFOR—Opposing Force

OTN—Own the Night

PEOC3S—Program Executive Officer for Command, Control, and Communications Systems

PEO—Program Executive Officer

PLGRS—Precision Lightweight GPS Receiver System

PLS—Palletized Loading System

PLT—Platoon

POS/NAV—Position/Navigation Initiative

PSSCS—Personnel Service Support Control System

RECON—Reconnaissance

RF—Radio Frequency

RWS—Remote Work Station

S/FOS—System of Systems and/or Federation of Systems

SDR—Surrogate Data Radio

SFIE—SOS and FOS Integration Environment

SINCGARS—Single Channel Ground and Airborne Radio System

SIP—System Improvement Program

SIV—System Integration Van

SOS—System of Systems

SPOEM—Special Program Office for Exploitation Modernization

STAMIS—Standard Army Management Information System

STM—System Test Mode

TELE-MED—Telemedicine

TF XXI—Task Force XXI

TMG—Tactical Multinet Gateway

TNS—Tactical Name Server

TOC—Tactical Operations Center

TPN—Tactical Packet Network

TRADOC—U.S. Army Training and Doctrine Command

TSIP—Task Force XXI System Integration Plan

UAV-SR—Unmanned Aerial Vehicle–Short Range

USDR&E—Under Secretary of Defense for Research and Engineering

USIGS—United States Imagery and Geospatial Information System

WAM—Wide Area Mine

X-FIST (BRAD)—Experimental Fire Support Team (Bradley)

Appendix D

Glossary of Terms

Several sources have been used for these frequently used terms. It is noted when adaptations have been made.

Architecture. The organizational structure of a system or component (IEEE Standard 610.12, 1990).

Battlefield digitization. The U.S. Army's program to apply information technologies to acquire, exchange, and use timely digital information tailored to the needs of each decider, shooter, and supporter (*Providing the Means,* 1994).

Core SOS. A minimum set of all unique hardware components and all software baselines of individual systems of the SOS and the architectural framework.

Coupling. Two systems are coupled if they are interdependent (i.e., if at least one system requires information from the other, or requires components, services, or people). Tighter coupling indicates greater interdependencies between systems than does loose coupling.

A federation of systems. A system of systems managed without centralized authority and direction.

Geospatial data. Information that identifies the geographic location and characteristics of natural or constructed features and borders of the earth (USIGS Glossary, 1998).

Heraclitan principle. A paraphrase of the sayings of the Greek philosopher, Heraclitus: "You can never experience the same SOS twice."

Information superiority. The capability to collect, process, and disseminate an uninterrupted flow of information while exploiting or denying an adversary's ability to do the same (Joint Vision 2010, 1996).

Infrastructure. The underlying processing, communications, and organization base, people, and processes, to support the function specified by the context in which the term is used.

Integration. The process of combining the components of a system into an overall system, or the process of combining the systems of a set of systems into a SOS (adapted from IEEE Standard 610.12-1990).

Integration environment. The people, common processes, and common infrastructure established at an integration site to support SOS (or FOS) integration. See Appendix B.

Integration testing. An orderly progression of testing in which the components of a system or systems of a set of systems are combined and tested until the system or set of systems has been evaluated (adapted from IEEE Standard 610.12-1990).

Interface. A shared boundary across which information is passed (IEEE Standard 610.12-1990).

The interfaces of a system. A connecting link or interrelationship between two systems, two devices, two applications, or the user and an application, device, or system (DISA DII Master Plan, 1998).

Interoperability. The ability of two or more systems or components to exchange information and to use the information that has been exchanged (IEEE Standard 610.12-1990).

Mapping, Charting, and Geodesy (MC&G). The collection, transformation, generation, dissemination, and storing of geodetic, geomagnetic, gravimetric, aeronautical, topographic, hydrographic, cultural, and toponymic data (USIGS Glossary, 1998).

Open system. A system that implements sufficient open specifications for interfaces, services, and supporting formats to enable properly engineered applications software: (1) to be ported with minimal changes across a wide range of systems; (2) to interoperate with other applications on local and remote systems; and (3) to interact with users in a style that facilitates user portability (DISA DII Master Plan).

Operational architecture. A description, often graphical, of the operational elements, assigned tasks, and information flows required to support the warfighter. It defines the type of information, the frequency of exchange, and what tasks are supported by these information exchanges (DISA DII Master Plan).

Shared situational awareness. The ability of a unit to know where its friends are located, where the enemy is, and to share that information with other friends, both horizontally and vertically, in near real-time (*Providing the Means,* 1994).

SOS cadre. The people responsible for architecting, engineering, testing, and integrating a SOS using systems managed, developed, and operated by (other) organizations and people.

System. A collection of components organized to accomplish a specific function or set of functions (IEEE Standard 610.12-1990).

System. A collection of different things that together produce results unachievable by themselves alone (Rechtin & Maier, 1997).

Systems architecture. A description that defines the physical connection, location, and identification of key nodes, circuits, networks, warfighting platforms, and so forth, and specifies system and component performance parameters. The systems architecture is constructed to satisfy operational architecture requirements per standards defined in the technical architecture. The systems architecture shows how multiple systems within a subject area link and interoperate and may describe the internal constructions or operations of particular systems within the architecture (DISA DII Master Plan).

A system of systems. A set of different systems so connected or related as to produce results unachievable by the individual systems alone.

Technical architecture. A description that identifies the services, interfaces, standards, and their relationships. It provides the technical guidelines for implementation of systems upon which engineering specifications are based, common building blocks are built, and product lines are developed (DISA DII Master Plan).

Appendix E

Bibliography

Advanced Battlespace Information System (ABIS) Task Force Report. Sponsored by Director for Command, Control, Communications, and Computers (Joint Staff) and Director of Defense Research and Engineering (OSD). May 1996.

Alberts, Davis S. *Operations Other Than War: The Technological Dimension*. Washington, DC: National Defense University Press. 1995. 3–4.

Army Technical Architecture, Version 4.5. At web site www.hqda.army.mil/webs/techarch/. 1996.

Arthur, W. Brian. "How Fast is Technology Evolving?" *Scientific American*. February 1997. 105, 107.

Barnett, Roger W. "Grasping 2010 with Naval Forces." *Joint Forces Quarterly,* 17. Autumn/ Winter 1997–1998. 25-31.

Baum, L. Frank. *The Wizard of Oz*. Illustrated Junior Library. New York: Grosset & Dunlap. First Copyright, 1900, 1903 by L. Frank Baum and W. W. Denslow.

Berry, B.J.L. "Cities as Systems within Systems of Cities." Papers of Regional Sciences Association, 13. 1964. 147–163.

Boehm, Barry, Bradford Clark, Ellis Horowitz, Chris Westland, Ray Madacy, and Richard Selby. "The Cocomo-2 Software Cost Estimation Model." *American Programmer,* IX:7. Cutter Information Corp. July 1996.

Boutelle, BG Steven, USA, and Alfred Grasso. "A Case Study: The Central Technical Support Facility." *Army RD&A.*, March–April 1998. 30–33.

Boutelle, COL Steven. TF XXI briefing slide. 1996.

Brownsword, Lisa, David Carney, and Tricia Oberndorf. "The Opportunities and Complexities of Applying Commercial-Off-the-Shelf Components." *Crosstalk: The Journal of Defense Software Engineering*. April 1998.

Butler, S., D. Diskin, N. Howes, and K. Jordan. "Architectural Design of a Common Operating Environment." *IEEE Software*, New York: Institute for Electrical and Electronics Engineers, 13:6. November 1996. 57–46.

Caldwell, BG John, USA. Interview. *Jane's Defence Weekly*. April 9, 1997. 32.

Coats, Dan. "Joint Experimentation—Unlocking the Promise of the Future." *Joint Forces Quarterly,* 17. Autumn/Winter 1997–1998. 13–19.

Cohen, William S. "The Secretary's Message." *Report of the Quadrennial Defense Review*. May 1997. At website http://www.defenselink.mil/pubs/qdr/msg.html.

Commission on Physical Sciences, Mathematics, and Applications. *Realizing the Potential of C4I: Fundamental Challenges*. National Research Council. Washington, DC: National Academy Press. 1999.

Concept for Future Joint Operations: Expanding Joint Vision 2010. Office of Primary Responsibility: Commander, Joint Warfighting Center, Fort Monroe, VA. May 1997.

Czerwinski, Thomas J. *Coping with the Bounds: Speculations on Nonlinearity in Military Affairs*. Washington, DC: National Defense University Press. 1998. Chaps. 1-2.

Czerwinski, Thomas J. "Command and Control at the Crossroads." *Parameters*, 26:3, Autumn 1996. 121–132. At website http://carlisle-www.army.mil/usawc/parameters/issues.htm.

de Laguna, Theodore. "The Importance of Heraclitus." *The Philosophical Review*, 30:3. May 1921. 238–254.

Defense Information Systems Agency. *Defense Information Infrastructure (DII) Master Plan*. Version 7.0. Glossary. March 1998.

Defense Systems Management College. "Best Practices and Lessons Learned." At web site www.dsmc.dsm.mil/.

Department of Defense Joint Technical Architecture (JTA), Version 2.0. At DISA web site, www-jta.itsi.disa.mil. May 1998.

Department of Defense Software Acquisition Best Practices Initiative. *The Program Manager's Guide to Software Acquisition Best Practices*. Version 2.0. April 1997.

Eisner, H. "RCASSE: Rapid Computer-Aided Systems of Systems (S2) Engineering." *Proceedings of the 3rd International Symposium of the National Council on System Engineering* (NCOSE), 1. 1993. 267–273.

Fox, Greg, Steven Marcom, and Karen W. Lantner. "A Software Development Process for COTS-Based Information Infrastructure." *Crosstalk: The Journal for Defense Engineering*. April 1998.

Gacek, Cristina and Barry Boehm. "Composing Components: How Does One Detect Potential Architectural Mismatches?" Workshop on Computational Architectures. Monterey, CA. January 6–8, 1998.

Gibish, Jane E. "Force XXI: A Selected Bibliography." U.S. Army War College Library. At web site http://carlisle-www.army.mil/library/bibs/forcexxi.htm. April 1997.

Goedkoop, COL Thomas R., USA, and CPT Barry E. Venable, USA. "Task Force XXI: An Overview." *Military Review*. March–April 1997.

Hagel, John. In an interview with Kevin Kelly. "It Takes a Village to Make a Mall." *Wired,* 5.08. August 1997. 4–6.

Hallion, Richard P. "Airpower and the Changing Nature of Warfare." *Joint Forces Quarterly,* 17. Autumn/ Winter 1997–1998. 39-46.

Handy, Charles. "Balancing Corporate Power: A New Federalist Paper." *Harvard Business Review*, 70:6. November–December 1992. 59–72.

Hanna, LTC Mark, USA. "Task Force XXI: The Army's Digital Experiment." National Defense University Strategic Forum, 119. Washington, DC: National Defense University Press. July 1997.

Hartzog, GEN William W., USA. "Task Force XXI Final Report Executive Summary." HQ TRADOC, DCSCD (ATCD-J) Force XXI/Joint Venture. Fort Monroe, VA. October 1997.

Hester, CPT Henry M., Jr., USA. "Digitization in Task Force XXI." *Field Artillery Journal*, September–October 1996. 42–44.

Hoffman, F.G. "Joint Vision 2010—A Marine Perspective." *Joint Forces Quarterly*, 17. Autumn/ Winter 1997–1998. 32-38.

Holcombe, Robert. "Some Lessons Learned While Digitizing the Battlefield." *Proceedings of the Battlefield Systems International 98 Conference.* June 9–11, 1998. 73–89.

Institute for Electrical and Electronics Engineers. IEEE Standard 610.12-1990, *IEEE Standard Glossary of Software Engineering Terminology.* New York. 1990.

"Joint Vision 2010, American's Military—Preparing for Tomorrow." *Joint Force Quarterly,* 12. Summer 1996. 34–49.

Joint Vision 2010. Office of Primary Responsibility: Chairman of the Joint Chiefs of Staff, The Pentagon, Washington, DC. 1996.

Joint Warfighting Science and Technology Plan. Department of Defense, Director of Defense Research and Engineering, Office of the Secretary of Defense, Washington, DC. February 1998.

Jones, Capers. "Software Project Management in the 21st Century." *American Programmer,* XI: 2. Cutter Information Corp. February 1998.

Jones, Capers. *Applied Software Measurement.* New York: McGraw-Hill. 1996a.

Jones, Capers. "Large Software System Failures and Successes." *American Programmer.* Cutter Information Corp. 1996b.

Jones, Capers. *Patterns of Software Systems Failure and Success*. International Thomson Press. 1996c.

Kelly, Kevin. "New Rules for the New Economy." *Wired*. September 1997.

Layton, Richard L. "Command and Control Structure." In Wentz, Larry, Ed. *Lessons From Bosnia: The IFOR Experience*. Washington, DC: National Defense University Press. 1998. 35–52.

Littlefield, Keith E. "The Defense Mapping Agency's Digital Production System (DPS)." *Cartography and Geographic Information Systems, Journal of American Congress on Surveying and Mapping*, 22:2. 1995. 119–127.

Maier, Mark W. "Architecting Principles for Systems-of-Systems." Preprint submitted to *Systems Engineering*. 1998.

Maier, Mark W. "On Architecting and Intelligent Transport Systems." *IEEE Transactions on Aerospace and Electronic Systems,* 3: 2. 1997. 610–662.

Maier, Mark W. "Architecting Principles for Systems-of-Systems." *6th Annual International Council on System Engineering (INCOSE) Symposium Proceedings*. 1996. 567–574.

Meigs, LTG Montgomery C., USA, and COL Edward J. Fitzgerald, III, USA. "University After Next." *Military Review*. March–April 1998. 37–45.

Naylor, Sean D. "Army Fires First Round in Battle for the Future." *Army Times*. April 28, 1997.

Owens, Adm William A., USN. "The Emerging U.S. System-of-Systems." *National Defense University Strategic Forum,* 63. Washington, DC: National Defense University Press. 1996.

Perrow, Charles. *Normal Accidents: Living with High Risk Technologies.* New York: Basic Books, Inc. 1984.

Providing the Means. U.S. Army Digitization Office. Washington, DC: Office of the Chief of Staff Army. 1994.

Rechtin, Eberhardt. *Systems Architecting: Creating and Building Complex Systems.* Englewood Cliffs, NJ: Prentice Hall. 1991.

Rechtin, Eberhardt and Mark W. Maier. *The Art of Systems Architecting*. CRC Press. 1997.

Reimer, GEN Dennis J., USA. "The Year in Review—The Year Ahead." Remarks at the Institute of Land Warfare. 1998.

Reimer, GEN Dennis J., USA. "Leaping Ahead to the 21st Century." *Joint Forces Quarterly*, 17. Autumn/Winter 1997–1998. 20-24.

Reimer, GEN Dennis J., USA. "Preparing Now to Meet 21st Century Challenges." Army Green Book, 47:10. October 1997.

Reimer, GEN Dennis J., USA. Dwight D. Eisenhower Speech. October 15, 1996.

Presidential Commission on Critical Infrastructure. *Critical Foundations: Protecting America's Infrastructures* (report).Washington, DC. 1997.

Royce, Walker. *Software Project Management: A Unified Framework.* Reading, Mass.:Addison-Wesley. 1998.

Royce, Winston. "Managing the Development of Large Software Systems." *Proceedings WESTCON,* California. 1970.

Ruocco, LTC Anthony, USA, and MAJ Todd L. Smith, USA. "Designing a Digital Library to Support Global Army Training." *Crosstalk: The Journal of Defense Software Engineering.* October 1998. 12.

Schaeffer, Mark. "Software Engineering and Systems Engineering in the Department of Defense." *Crosstalk: The Journal of Defense Software Engineering,* 11:10. October 1998. 3–6.

Shelton, GEN Henry H., USA. "A Word from the New Chairman." *Joint Forces Quarterly*, 17. Autumn/Winter 1997–1998. 6-8.

Shelton, GEN Henry H., USA. "Remarks at the General Graves B. Erskine Distinguished Lecture Series." Quantico, VA: Marine Corps University. February 23, 1998.

Shenhar, A. "A New Systems Engineering Taxonomy." *Proceedings of the 4th International Symposium of the National Council on System Engineering*, 2. 1994. 261–276.

Sheth, Amit P. and James A. Larson. "Federated Database Systems for Managing Distributed, Heterogeneous, and Autonomous Databases." *ACM Computing Surveys*, 22: 3. 1990. 183–236.

Slabodkin, Gregory. "First-Blush Review of Force XXI Finds Army Training Inadequate." *Government Computer News*. 1997.

Stanley, Elizabeth A. *Evolutionary Technology in the Current Revolution in Military Affairs: Army Tactical Command and Control System.* Strategic Studies Institute, U.S. Army War College. 1998.

Starr, Stuart H. "Perspectives on C4I Interoperability." *Proceedings of the Naval War College, Command and Control Research and Technology Symposium*. Monterey, CA: Naval Post Graduate School. 1996. 331–345.

Stutzke, Richard D. "Software Estimating Technology." At web site www.saic.com/publications/.

Sullivan, GEN Gordon R., USA. Memorandum, March 9, 1992. Re: Louisiana Maneuvers 1994. In Sullivan and Coroalles. *Seeing the Elephant: Leading America's Army into the Twenty-first Century*. Cambridge, MA: Institute for Foreign Policy Analysis. 1995. 55–56.

Sullivan, GEN Gordon R., USA, and COL Anthony M. Coroalles, USA. *The Army in the Information Age*. Strategic Studies Institute, U.S. Army War College. 1995.

Sullivan, GEN Gordon R., USA, and COL James M. Dubik, USA. *War in the Information Age*. Strategic Studies Institute, U.S. Army War College. 1994.

Task Force XXI Architecture. At web site www.monmouth.army.mil/cecom/lrc/forcexxi/equipa.html.

Task Force XXI AWE Integrated Report Annotated Briefing Slides. U.S. Army Training and Doctrine Command (TRADOC). 1997.

Task Force XXI outbrief to the U.S. Army Training and Doctrine Command (TRADOC) staff. At web site www.monroe.army.mil/pao/tradoc/sld001-sld021.html.

Task Force XXI System Integration Plan (TSIP). 1996.

The Standish Group, CHAOS survey. At web site www.standishgroup.com. 1995.

The White House, *A National Security Strategy for a New Century*. Washington, DC. 1998.

United States Imagery and Geospatial Information System (USIGS) Glossary, Revision A. National Imagery and Mapping Agency (NIMA). 1998.

Wentz, Larry, Ed. *Lessons From Bosnia: The IFOR Experience*. Washington, DC: National Defense University Press. 1998.

Wright, Randall R. "Process Standards and Capability Models for Engineering Software-Intensive Systems." *Crosstalk, The Journal of Defense Software Engineering*, 11:10. October 1998. 7–11.

U.S. GOVERNMENT PRINTING OFFICE: 1999 -- 454 - 121 / 00007

Order Form

The publications listed on this form are available at no cost through CCRP. Simply mark the publication(s) you would like to receive and return these sheets to the following address:

CCRP Publications Distribution Center
c/o Evidence Based Research, Inc.
1595 Spring Hill Road, Suite 250
Vienna, VA 22182-2216

or fax to:

(703) 821-7742

- ❑ Please change my listing in your database.
- ❑ Please add my name to your database.
- ❑ Please remove my name from your mailing list.

Name ______________________________
Title ______________________________
Company ______________________________
Address ______________________________
City ____________ State ______ Zip ________ Country ____
Phone ______________________________
E-mail ______________________________
Area(s) of Expertise ______________________________
I would like to see more publications on the following subject(s):

____

I use these publications primarily for the following reason(s):

- ❑ Personal reference
- ❑ Work reference
- ❑ Course textbook

For further information on CCRP, please visit our website at
www.dodccrp.org

Command and Control

- ❑ Command, Control, and the Common Defense (Allard)
- ❑ Command and Control in Peace Operations Workshop
- ❑ Command Arrangements for Peace Operations (Alberts & Hayes)*
- ❑ Complexity, Global Politics, and National Security (Alberts & Czerwinski, eds.)
- ❑ Coping with the Bounds (Czerwinski)

Information Technologies and Information Warfare

- ❑ Behind the Wizard's Curtain: An Integration Environment for a System of Systems (Krygiel)
- ❑ Defending Cyberspace and Other Metaphors (Libicki)*
- ❑ Defensive Information Warfare (Alberts)*
- ❑ Dominant Battlespace Knowledge (Johnson & Libicki)*
- ❑ The Information Age: An Anthology on Its Impacts and Consequences (Papp & Alberts, eds.)*
- ❑ Information Warfare and International Law (Greenberg, Goodman, & Soo Hoo)
- ❑ The Mesh and the Net: Speculations on Armed Conflict in a Time of Free Silicon (Libicki)
- ❑ Network Centric Warfare: Developing and Leveraging Information Superiority (Alberts, Garstka, & Stein)
- ❑ Standards: The Rough Road to the Common Byte (Libicki)*
- ❑ The Unintended Consequences of Information Age Technologies (Alberts)*
- ❑ What is Information Warfare? (Libicki)*

Operations Other Than War

- ❑ Confrontation Analysis: How to Win Operations Other Than War (Howard)
- ❑ Doing Windows: Non-Traditional Military Responses to Complex Emergencies (B. Hayes & Sands)
- ❑ Humanitarian Assistance and Disaster Relief in the Next Century (Sovereign)
- ❑ Interagency and Political-Military Dimensions of Peace Operations: Haiti—A Case Study (Daly Hayes & Wheatley, eds.)
- ❑ Lessons from Bosnia: The IFOR Experience (Wentz, ed.)
- ❑ NGOs and the Military in the Interagency Process (Davidson, Daly Hayes, & Landon)
- ❑ Operations Other Than War*
- ❑ Shock and Awe: Achieving Rapid Dominance (Ullman & Wade)*
- ❑ Target Bosnia: Integrating Information Activities in Peace Operations (Siegel)

Proceedings of the Command and Control Research and Technology Symposium (CD-ROM Only)

- ❑ 1998/4th International
- ❑ 1999

*Published in conjunction with the NDU Press.